NATURE WARS

NATURE WARS

People vs. Pests

Mark L. Winston

HARVARD UNIVERSITY PRESS
Cambridge, Massachusetts
London, England

Printed in the United States of America
Third printing, 1999

First Harvard University Press paperback edition, 1999

Library of Congress Cataloging-in-Publication Data
Winston, Mark L.
Nature wars : people vs. pests / Mark L. Winston.
p. cm.
Includes bibliographical references (p.) and index.
ISBN 0-674-60541-1 (cloth)
ISBN 0-674-60542-X (pbk.)
1. Pesticides—Environmental aspects.
2. Insect pests—Biological control.
3. Nature—Effect of human beings on. I. Title.
QH545.P4W55 1997
577.27'9—dc21
97-17302

Contents

Preface

Every kind of organism has some defining characteristics by which it can be identified as an entity different from all others. A piercing howl on a cold winter night says "I am wolf." The annoying hum of a mosquito, the soaring trunk and fine-smelling wood of a cedar tree, the lumbering walk of a bear—these, too, are shorthand signals that announce membership in a particular group. We humans also have signature characteristics, and it must be simple for other animals to recognize us by our bipedal posture, front-facing eyes, and relatively hairless bodies. The most unusual aspect of being human, however, may be the degree to which we consider our existence and ponder how we fit into a world that is becoming increasingly distant from the environments where we evolved.

We are an ambiguous species, living simultaneously in and outside of nature. We are subject to the same biological laws as other animals—we age, our populations outgrow their food supplies, we compete for mates. But we are unlike other species on earth in considering ourselves separate from, and more important than, the rest of life. The earliest recorded human musings about nature exhibited this distance, and contemporary thought maintains a sense of separation from the nonreflective natural world.

This sense of distance is at the core of the environmental crisis we find ourselves in today. Perhaps the most disturbing aspect of our negative impact on the rest of nature is not that it is happening but that we are doing it consciously. We have allowed our sense of dominion over the rest of the earth to justify remodeling the globe to suit our needs. But global renovation has come at a cost, and is having effects on ourselves and other species that we are only beginning to comprehend.

These human quirks of distancing ourselves from nature and elevating our significance above other species might never have become important in the larger scheme of things except for the rapid advances in science and technology that have shaped recent human history. Where early humans may have carved a few fields out of the forest or prairie, contemporary humans have left only a few token patches of forest and prairie in a sea of grain. Our ancestors burned a trivial amount of wood for heat, whereas today we burn so much fuel that we are modifying the earth's climate. A protective ozone layer that took billions of years to develop has been severely diminished in just a few decades by our extensive use of ozone-depleting chemicals in aerosol spray cans and refrigerators. The dirt trails where our ancestors once walked have become asphalt superhighways, and our automobiles daily pollute the atmosphere with tons of exhaust, to the point where we can no longer safely breathe the air in many of our cities. Our ability to radically transform the world has caught up with our historical, human-centered sense of dominance and distance. Although we are unlikely ever to return to a time when we were "natural," our self-interest is gradually forcing us to develop a perspective of stewardship rather than dominion if we wish to survive. Today's growing environmental ethic exists because our impact on the globe is finally fouling our own nest.

This book reflects my own attempt to understand one aspect of our contemporary human relationship with nature: how we deal with pests. I first became aware of pests in the early 1970s, as part of a summer job working for the Gypsy Moth Methods and Development Section of the U.S. Department of Agriculture on Cape Cod. Like most summer jobs, it was boringly routine. It involved driving the dirt roads of Cape Cod in a Jeep looking for gypsy moth infestations that the USDA biologists could use to test various control sprays. I have two vivid memories from that summer. The first was of wandering through defoliated forests, smelling the stench of rotting gypsy moth larvae that had died from starvation and viral diseases after consuming every leaf in an outbreak area. I also recall putting out plastic ribbons to demarcate spray zones for test programs, and coming back the next day to find that the ribbons had been impressively consumed by the voracious moths.

That particular job convinced me that I needed more education, and I went on to study entomology. In 1980 I joined the faculty at the Centre for Pest Management, a subsection of the Department of Biological Sciences at Simon Fraser University in Vancouver. There, I conducted research on honey bee behavior and management, an unusual role in the Centre because I do not study a pest. The honey bee is a beneficial insect, which provides a nice balance to the pestiferous weeds, insects, and fungi that other Centre members study how to kill, albeit in environmentally friendly ways. Research in our Centre is "progressive" and has little to do with pesticides. Rather, we focus largely on alternative management methods involving pheromones, parasitic insects, and bioengineered crops, or pest diseases that do not have the side effects of synthetic chemicals.

It was my reacquaintance in Vancouver with the gypsy moth that led to this book. In the early 1990s the city of Vancouver became infested with gypsy moths that arrived simultaneously from eastern Canada and Siberia. Agriculture Canada (the Canadian equivalent of the USDA) decided to spray the city with a biological pesticide, the bacterium *Bacillus thuringiensis* (see Chapter 2 for a full account of this program). My colleagues and I in the Centre approved, and then were stunned to discover that the citizens of Vancouver did not share our enthusiasm for being sprayed with bacteria, no matter how much we assured them of its safety and environmental correctness.

I began writing this book in response to that incident, partly with the desire to explain these new biological methods, but more broadly out of a developing curiosity to explore the factors that influence our decisions about how to deal with pests. Because of my experience in the Centre for Pest Management, I expected to find a rapidly diminishing use of pesticides, replaced by a cornucopia of novel biologically based strategies that were becoming standard practice. That was not what I found. I was surprised to discover that most biologically based methods remain at the fringes of pest management, in the realm of potential but unproven techniques. Alternative technologies, while scientifically very exciting, in practice only nibble at the edges of chemical pesticides, which still dominate the world of pest management—in our cities and homes, our fields and forests, our parks and lawns. My focus changed as this book developed, and became less exuberant about these wonderful alternatives and more involved with understanding why biologically based methods have not been more widely adopted.

Pests and how we manage them are interesting in their technical complexity but also expressive of how we humans consider our place in nature. We rarely look beyond the immediate benefits and disadvantages, costs and side effects, of pest-managing methods. Yet the choices we make in dealing with pests reflect our underlying values as individuals and as a society concerning the natural world. To date, and in spite of considerable environmental protest, public outcry, and viable alternatives, we still choose to spray chemical pesticides when faced with a pest management decision.

I am not opposed to the use of chemicals against pests when required, but I believe we can do considerably better. My hope is that this book will explain in part why we remain rooted in chemically based methods, and that this understanding will lead us to begin choosing alternatives that are more environmentally compatible and less harmful to our own health.

I have chosen to investigate this subject in the following chapters through various American and Canadian case histories. While each describes a unique situation when viewed separately, these chapters taken together reveal a pattern in our basic concept of what constitutes a pest and how to deal with those organisms that do us harm. Currently, we approach pests as organisms to control rather than manage; we exterminate instead of reduce; we dominate rather than learn to accommodate. Pest management has become a modern war against nature, but it is time to reconsider the terms of engagement.

NATURE WARS

CHAPTER ONE

A Pestiferous World

"All this is not to say there is no insect problem and no need of control. I am saying, rather, that control must be geared to realities, not to mythical situations, and that the methods employed must be such that they do not destroy us along with the insects."

Rachel Carson, *Silent Spring* (1962)

Humans, like all other organisms, have competitors for our food and shelter, predators that attempt to eat us, and parasites that feed in or on us. We have designated the species that compete with us in these natural associations with a special term, "pests." Perhaps it is appropriate, if somewhat ironic, to use this specialized terminology for interactions which, when they occur in other organisms, we consider perfectly natural and right, because many pest problems that afflict humans are of our own making. Our human activities have exacerbated pest situations far beyond their natural significance. Thus, pests provide an excellent paradigm to understand how our attempts to have dominion over nature have backfired, and a perspective to us in developing new attitudes about nature that might lead us out of the environmental problems we have created.

Our reactions and overreactions to pests demonstrate an ancient realization that human survival is directly related to our ability to manage pests. An extraordinary range of organisms plague people, and they have impact on virtually every aspect of our capacity to survive and reproduce.

Rats, mice, cockroaches, termites, beetles, moths, ants, lice, fleas, mosquitoes, spiders, mites, ticks, pigeons, raccoons, coyotes, deer, woodchucks, nematodes, fungi, weeds, and myriad other organisms in a bewildering diversity of forms infest human life. Pests eat our food, consume our homes, transmit or cause human diseases, devour our clothing, and are often excruciatingly irritating when they inhabit our skin, hair, or digestive tract. Pest control is not trivial; human life itself is often threatened by pests, and it is appropriate that our attitudes about pests often take on a kill or be killed perspective.

The opposing perspective, live and let live, does not usually arise when considering pests. We do not have warm and fuzzy feelings about pests, only hard action focused on eradicating them, or at least reducing pest populations to the point where they do little damage. Given the damage pests can cause, it is not surprising that there are few advocates for pests, and that our thoughts about them tend toward the pragmatic rather than the philosophical.

Nevertheless, our survival-driven need to control pests has led to serious environmental issues that have arisen as a result of pest management, issues that affect both humans and the rest of the earth's inhabitants. Our extensive colonization of the earth's habitats has made pests out of many previously innocuous organisms or worsened the impact of formerly minor pests. In addition, pests frequently develop resistance to our chemically driven pest management techniques, and we have been forced into an escalating arms race to develop new tools to deal with increasingly virulent versions of pests that we have created by our management strategies.

Pest management decisions today range from the highly local to the global, and each decision has impact well beyond the immediate pest problem. Each time we purchase a can of Diazinon insecticide to kill kitchen ants, or spray a squirt of Roundup herbicide on a dandelion, we become a small cog in an extensive management system built out of chemical pesticides. When we spray a corn field with Sevin, or an apple orchard with Guthion, the drift from that spray ends up in tree bark at the other side of the world, or in lichens well north of the Arctic Circle. Every rat we kill with anticoagulant pesticides, every gypsy moth that dies from an aerial spray, or raccoon that is trapped while consuming garbage in our cities represents a decision we as a society have made about our environmental values and management paradigms. But most pest control is largely hidden from the direct sight of individuals, surfacing only in occasional public battles between environmentalists and the commercial pest management industries.

Today, decisions concerning how to manage pests have become a major focus of the environmental agenda. The issues have become complex, with political, economic, sociological, and personal factors having considerably more weight than questions of biological impact. At the core of pest management decisions lies the contradiction that frequently confronts us when we consider implementing any scientific advance that has the potential to improve human life, but also can have side effects. Our pest management paradox is that the most commonly prescribed treatment, chemical pesticides, can be damaging to ourselves and our environment. 

❧

The history of pests and humans is long, but the impact of pests and their management has intensified as our role on earth has become more dominant and encompassing. Life for our early ancestors was relatively free of pests. The earliest hominids did not grow or store food and had no permanent homes. Their pests were limited to biting insects such as lice, fleas, flies, and mosquitoes. Pest control techniques were simple: scratching, grooming, swatting, and squashing.

Beginning around 10,000 years ago, with the advent of agriculture and denser human settlements, our relationship to pests began to change. Increasing urban populations, compact crop plantings in fields and orchards, herds of domesticated animals, and stores of grains, vegetables, cloth fibers, furs, and dried meats all provided concentrated food sources for organisms that previously foraged widely for sparse food sources. In addition, we began to disrupt entire ecosystems and their inhabitants, transforming diverse natural habitats into cultivated, single-cropped fields and dense, sprawling cities. These changes induced the populations of a small number of species to explode into pest status. Trade added to this potent brew by transporting plants and animals far out of their natural ranges, to new habitats with exciting food sources and few predators, parasites, or diseases to keep them in check.

Pests had become a major problem by the time of the earliest recorded human history, and pest management occupied a considerable part of early society's attention. Biblical writings refer frequently to plagues of locusts; the inclusion of locusts in the Old Testament as kosher animals acceptable for human food is thought to have served as both a pest control measure and a dietary substitute for the grain their swarms consumed. Remains of grain-eating beetles have been found in vases in King Tutankhamen's tomb dating back to about 1350 B.C. The Romans had a special Latin term, *muricidus,* for professional rodent exterminators. In the New World, pre-Columbian pottery depicts swarms of flies.

Pest management in these early times was erratic, involving religious ceremonies and superstitious practices, certainly with very limited success. Pest control was the province of priests and witch doctors who employed magical chants, potions, and prayers to alleviate afflictions of the body, crops, and stored foods. The Greeks assigned particular gods to prevent or exterminate vermin; Apollo had domain over mice and mildew, Hercules was invoked against locusts and worms, and all-powerful Zeus was nicknamed the "Fly-catcher." The Romans also appointed various deities for pest control, such as Robigus, the god responsible for control of cereal fungal rusts, and even conducted a special annual festival called the Robigalia in which they sacrificed a red puppy to appease Robigus.

Even the new religion of Christianity developed rituals to control pests that persisted until close to modern times. As late as the fifteenth century, in Berne, Switzerland, insect cutworms infesting crops were tried in religious court, found guilty, excommunicated, and banished. In 1545 a similar trial was held in the French village of Saint-Julian, but with a different outcome. The town's grape growers asked the ecclesiastical authorities to excommunicate weevils infesting their vines, but the counsel appointed to defend the weevils argued the unusual position that weevils, like man, were created by God and had equal rights to the grapes. The judge ruled in favor of the weevils.

Not all early pest control techniques involved deities. Pest managers recommended various "procedures" that were considered important in pest control. The ancient Syrians exorcised scorpions from their capital city of Antioch by burying a bronze figure under a small pillar at the city's center. Pliny the Elder reported in the first century B.C. that naked menstruating women walking around grain fields could force caterpillars and beetles infesting the plants to fall dead to the ground. The Romans also believed that staking the skull of an ass or mare in the garden would provide protection against caterpillars. Farmers in Elizabethan Ireland attempted to rhyme rats to death, perhaps with bad puns, while in Scotland bagpipes were used to drive rats away.

Early pest management was not all superstition and puppy sacrifices. The roots of many of our current nonchemical pest management techniques can be found in early pest control practices. For example, the widespread Roman practice of draining swamps, building sewage removal systems, and using baths contributed to improved sanitation and diminished pest problems. Another Roman contribution to pest control was made by the architect Marcus Pollio, who designed a granary made of granite slabs and wood that rested on top of circular stone rat guards, to prevent rats from climbing into the grain storage chamber. Pliny recommended storing grain in *siri,* airtight chambers in which carbon dioxide built up from respiring seeds, creating an environment inhospitable to most grain-eating insects. The Greeks also made some advances in pest management. Homer recognized the value of burning fields to control locusts, and in 450 B.C. Herodotus reported the use of mosquito netting and high sleeping towers to prevent mosquito bites.

The forerunners of today's chemical pesticides also appeared early in human history. The first recorded use of a pesticide was by the Sumerians, who were using elemental sulfur to control insects and mites by 2500 B.C. This practice was continued and enhanced by the Romans, who added oil to the mixture and used it as an insect repellent. The Romans introduced a number of new chemicals, including a fumigant created by boiling olive oil lees, bitumen, and sulfur that was used to protect vines from caterpillars. Both Greek and Roman agriculturists recommended soaking seeds in leek or cucumber extracts before planting.

The Greek agricultural encyclopedia *Geoponika* noted that bay, asafetida, elder, cumin, hellebore, oak, squill, cedar, absinthe, garlic, and pomegranate all were effective as insect control agents. The *Geoponika* also reported the first pest strip, a mixture of bay and black hellebore in milk or sweet wine that was both attractive and fatal to flies. All of these plants and their extracts were later found by modern chemists to contain insecticidal agents.

Innovations in pest management were not confined to Mediterranean cultures. Pest management by the early Chinese showed a high degree of sophistication growing out of their long experience rearing silkworm moths. Chinese pest management techniques included using herbs, oils, chalk, and ash to protect seeds and stored grain, applying mercury and arsenic compounds to control body lice and other pests, dipping rice roots in white arsenic to protect transplanted rice from insect damage, and protecting sheep from various parasites by using pig oil. The Chinese also deserve credit for the earliest recorded biological control measure, in which colonies of predatory ants were established in citrus orchards to suppress caterpillars and beetles. Bamboo bridges were tied between branches and trees to facilitate ant movement.

These early advances in pest control halted abruptly about the time the Roman Empire dissolved. Neither Eastern nor Western cultures made many scientific advances in the ensuing centuries, and it was not until the Scientific Revolution that science and pest control again began to make progress.

❧

A large part of the renewed interest in pest control was the direct result of the dramatic growth of agriculture. In the 1700s the change in food production from a subsistence to a commercial enterprise introduced such practices as extensive manure and fertilizer applications, expanded acreages, and row planting. The use of machines to seed, weed, fertilize, harvest, store, transport, and process food accelerated this growth. These changes in agriculture critically increased the number of pest problems, to the point that by the mid-1800s European countries and their colonies experienced significant pest-related agricultural disasters, including the total loss of many crops. The potato blight in Ireland, England, and Belgium, powdery mildew in European grapes, fungus leaf spot coffee disease in Ceylon, which forced the switch from coffee production to tea, and the grape phylloxera insect, which almost destroyed the French wine industry, are just a few of the agricultural problems that could be laid at the feet of pests.

The severity of these pest problems was a consequence of two major factors. First, the planting of vast acreages in one or only a few crops provided an excellent source of food for any organism that could feed on that particular plant. Consequently, reproduction among those organisms, no longer checked by scarce food supplies, spiraled out of control. Second, imperial colonization and increased international trade transported organisms to new environments around the world, where their proliferation in the presence of new food sources and in the absence of predators turned them into pests.

At about the same time that the expansion of agriculture and trade were producing new categories of pests, medical scientists were becoming aware that the infectious organisms responsible for many epidemics in human populations are carried (vectored) by pests. The first demonstrated case was not a human disease but an outbreak of Texas cattle fever in 1893, in which ticks were shown to carry and transmit the one-celled organism that caused the disease. This finding quickly led to the discovery of other pest vectors, including tsetse flies carrying the African sleeping sickness pathogen, rat fleas harboring plague, and mosquitoes vectoring the malaria parasite and yellow fever virus. The awareness that pests vectored human and animal pathogens and parasites stimulated the development of chemical control measures for the vector organisms.

Our modern chemical pesticide industry can be traced to an accidental observation by a grape farmer in Europe whose concern was the human "pests" who were stealing his grapes. He sprayed a poisonous-looking mixture of copper and lime at the edge of his grape fields to dissuade pilferers, but later observed that these edge plants had escaped infection by the fungus that causes powdery mildew. This observation led to the development of the Bordeaux mixture, containing hydrated lime plus copper sulfate, which even today is the most commonly used fungicide in the world.

The French grape growers also began using a mixture of copper acetoarsenite, called Paris Green, that proved to protect their plants against damage by various insects. In the ensuing years, other inorganic compounds composed primarily of arsenic, antimony, selenium, sulfur, thallium, zinc, and copper were developed and used widely in pest management. The invention of a simple mechanical device, the spray nozzle, and the development of airplanes and airborne spraying techniques completed the early stages leading to today's chemically based pest management.

The onset of World War II moved us into the modern era of synthetic organic pesticides. Troops and civilians in Europe during World War I had suffered from lice, fleas, and bed bugs, and casualties from diseases vectored by these insects were considerable. But World War II was another story altogether. It was fought largely in subtropical and tropical areas in Africa and Asia, where insect-vectored diseases such as malaria, sleeping sickness, dengue, and relapsing fever had the potential to devastate the war effort. In addition, powerful chemical warfare agents were being investigated that could debilitate or kill enemy troops in large numbers. The chemistry of these two types of "weapons"—against insects and human enemies—involved similar compounds and, used together, could potentially be decisive for the side that managed to develop them first.

Both the Allied and Axis powers screened and tested hundreds of compounds for insecticidal and antipersonnel activity, and each made breakthroughs that led to the development of today's insecticides. The first major advance occurred in the United States, where a compound being tested by the Swiss chemist Paul Mueller proved highly toxic to virtually all insects, and at extremely low doses. This compound was dichloro-diphenyl-trichloroethane, commonly known as DDT. It is classified as a chlorinated hydrocarbon pesticide, and it and other closely related compounds did prove to have significant wartime benefits in diminishing vector-carried casualties.

The military interest in using DDT and similar chemicals not just to destroy pests but to poison enemy soldiers is due to the mode of action of most insecticidal compounds. Insecticides generally act on the nervous system, either by narcotic action or by interrupting the transmission of nervous signals in various ways. Unfortunately, the nervous systems of insects and vertebrates are basically similar, so that most insecticides are toxic to people as well as insects. Insects, however, are much smaller than people, and it takes a considerably lower dose of insecticide to poison an insect than it does to poison us. Nevertheless, synthetic and botanically derived insecticides can be toxic to humans and other vertebrates, depending on the dose and mode of exposure.

The advent of synthetic pesticides revolutionized agriculture after the war and led to farming practices that became highly dependent on chemical use. Previously, farmers had attempted merely to control pests, but the post-World War II generation focused on complete pest eradication. The new chemicals were so effective, and so cheap, that nonchemical pest management techniques such as rotating crops, removing diseased plants or animals, encouraging natural enemies, draining standing water, and selecting resistant varieties were no longer practiced. Spraying was conducted on a schedule rather than by pest assessments, and pesticide companies became the farmer's primary source of information about pest management.

Although these chemicals have many benefits, they also have a cost beyond their purchase price when overused, improperly handled, and poorly applied. A groundswell of public antagonism against synthetic pesticides developed during the 1950s, as evidence mounted that pesticides such as DDT and others were responsible for massive kills of fish, birds, plants, and other organisms, as well as threatening human health. The environmentalist's response to the growing and indiscriminate use of chemical pesticides soon found its most eloquent champion with the publication of *Silent Spring* by Rachel Carson in 1962.

The defining sound-bites for pest management in our century are the phrases "Rachel Carson" and "DDT." Her name and those initials have come to represent two opposing viewpoints that still define the essence of our contemporary dilemma concerning the impact of technology on our lives. Carson was the quintessential defender of the environment, a champion of human health and ecological wellness. DDT, in contrast, represented modern science in service to humanity; DDT—the product of human ingenuity, and backed up with all the public relations hype and promise that industry and agriculture could deliver—became the flagship chemical in our ongoing war to demonstrate our dominion over the rest of nature.

Carson's message was not nearly as extreme as her detractors made it out to be. She was not a raving, chemophobic radical, but rather had a balanced ecological vision of how we could both manage pests and protect the environment. She also was not completely antichemical, but believed that pesticides should be used sparingly, as a last resort, and only following extensive investigations concerning their impact on non-target organisms. She wrote in *Silent Spring:* "It is not my contention that chemical insecticides must never be used. I do contend that we have put poisonous and biologically potent chemicals indiscriminately into the hands of persons largely or wholly ignorant of their potentials for harm . . . I contend, furthermore, that we have allowed these chemicals to be used with little or no advance investigation of their effect on soil, water, wildlife, and man himself. Future generations are unlikely to condone our lack of prudent concern for the integrity of the natural world."

Her message appears reasonable in light of our current understanding of chemical pesticides and the environment, but when first published her book whipped up a firestorm of extreme reactions, often from people who had not read it. The *Globe-Times* newspaper in the farming community of Bethlehem, Pennsylvania, wrote: "No one in either county farm office who was talked to today had read the book, but all disapproved of it heartily." One well-known and often quoted comment about Carson came from a meeting of the Federal Pest Control Review Board: "I thought she was a spinster. What's she so worried about genetics for?" A pre-publication lawsuit threat to the publisher suggested that Houghton-Mifflin "might wish to reconsider its plans to publish *Silent Spring,* especially in view of the book's inaccurate and disparaging statements about chlordane and heptachlor" (two pesticides closely related to DDT). The Director of the Montrose Chemical Company, a manufacturer of DDT, said that Carson wrote not "as a scientist but rather as a fanatic defender of the cult of the balance of nature . . . As science, *Silent Spring* is so much hogwash."

Opposition to Carson was based on two perspectives. First and most obvious was that a reduction in pesticide use would reduce industrial and agricultural profits. The Director of the New Jersey Department of Agriculture wrote that "in any large scale pest control program we are immediately confronted with the objection of a vociferous, misinformed group of nature-balancing, organic gardening, bird-loving, unreasonable citizenry that has not been convinced of the important place of agricultural chemicals in our economy." The National Agricultural Chemicals Association also focused its attacks on economic issues: "A serious threat to the continued supply of wholesome, nutritious food, and its availability at present-day low prices is manifested in the fear complex building up as a result of recent unfounded, sensational publicity with respect to agricultural chemicals."

The vociferous response to *Silent Spring* also reflected different visions of technology's role in improving human life. Carson's book was published at a time when we were beginning to realize that scientific advances were double-edged swords, with potential both to improve our lives and to damage ourselves and our environment. This was an era when technology was backfiring, with nuclear testing, thalidomide, DDT, and many other innovations wreaking havoc on an increasingly concerned public. Our ability to control the harmful side of science had not kept pace with the potential of science and technology to improve human life and protect the environment.

Today that balance is still an issue, and the two sides continue to attack and counterattack with extremes of language. Polarized debates go on, with chemicals such as Agent Orange, Dimilin, and Alar taking the place of DDT. However, the issues today are more complex and subtle. Extensive regulations appear to protect us from the most toxic chemicals but lull us into a false sense of complacency that our use of chemical pesticides has diminished.

Not so. The extent and impact of our current dependence on pesticides for both agricultural and nonagricultural purposes is staggering. In 1993, 1.1 billion pounds of active pesticide ingredients were used in the United States, and 4.5 billion pounds world-wide. The U.S. figure translates to about 4 pounds of pesticides for every man, woman, and child in America. Considering that toxic dosages of most pesticides to humans are about one hundred thousandth to one millionth of a pound, that's a considerable amount of poison. The pesticide industry in the United States was valued at $8.5 billion that year, with three-fourths of that value going toward agricultural pesticides. There were 860 active ingredients registered for use under the Federal Pesticide Act, formulated in 21,000 different products manufactured by 1,200 producers, and applied by 1.3 million certified pesticide applicators. These chemicals were used on 900,000 farms and in 69 million households in 1993.

Our commitment to pesticide-based pest management has led to an escalation of pests on our farms, in our forests, and throughout our homes and businesses, owing to pest resistance and the elimination of natural predators and parasites. It has also resulted in environmental contamination and numerous human health problems. In retrospect, this is certainly not surprising; spraying a billion pounds of anything across the United States is bound to have side effects, let alone spraying that much of a potent chemical.

The statistics concerning the impact of pesticides on human health and the environment are numerically overwhelming. A 1989 World Health Organization report estimated that there are about one million cases of human pesticide poisoning world-wide annually, resulting in 20,000 fatalities, mostly in developing countries where regulations concerning pesticide use are looser than in North America. Chronic pesticide exposure among humans has been linked to infertility, immune dysfunction, mood disorders, and various forms of cancer and birth defects. A 1993 estimate by David Pimentel of Cornell University put the cost of human pesticide poisonings and pesticide-related illness at $787 million annually in the United States, with $707 million of that attributed to treatment costs for an estimated 10,000 pesticide-related cancer victims. These data are based on epidemiological evidence showing a significantly higher incidence of cancer in U.S. farm workers than in the general public, possibly associated with the high frequency with which they use certain pesticides. This type of epidemiological data is inferential and does not prove that pesticides induced each case. Nevertheless, they suggest some concern about the carcinogenicity of pesticide use, at least for pesticide applicators.

Pesticides have other costs to human society in addition to medical expenses. For example, American losses from livestock fatalities and destruction of pesticide-contaminated animal products such as milk, meat, and eggs are about $30 million annually. Decontamination is an even bigger-ticket item; monitoring and clean-up of groundwater polluted by pesticides costs $1.8 billion annually in the United States. The removal of pesticides from groundwater and soil surrounding only one contaminated pesticide storage site, the Rocky Mountain Arsenal near Denver, cost $2 billion in the late 1980s.

Pimentel and colleagues estimate that the economic impact from the side effects of pesticide use in the United States is about $8.1 billion each year, based on public health costs, loss of domestic animals and their products, fish losses, monitoring and cleanup expenses, cost of government regulations, and deleterious impact of pesticides on predators and parasites of pests. If we add to this figure the $8.5 billion value of the pesticide industry, then the annual pesticide budget in the United States comes to about $17 billion. To put this value in some perspective, Pimentel estimated that pesticide use saves about $16 billion in crops annually from pests. When other human benefits of pesticide use are considered, such as vector control, pesticide use may show a slight positive economic return, but it would be difficult to argue for such extensive pesticide use on purely financial grounds.

These economic considerations do not include the impact of pesticides on nonhuman, nontarget species. For example, Pimentel cites data suggesting that pesticides kill about 67 million birds and between 6 and 14 million fish each year in the United States. Fish are especially vulnerable because pesticide runoff spreads quickly through aquatic environments, while birds are susceptible since commonly used food items such as grain and fish biomagnify pesticides, leading to chronic and acute poisoning after they are consumed. Pesticides are ubiquitous in virtually all habitats, even far from sprayed areas decades after spraying, and have negative impact on everything from soil-dwelling microorganisms to plants to beneficial insects, in addition to vertebrates.

Over the long run, pesticides frequently lead to the evolution of resistant pests, particularly among insects, which reproduce frequently and have high birth rates, allowing surviving populations to respond quickly following pesticide applications. Also, insects have enzyme systems that originally may have evolved to detoxify chemical defenses that occur naturally among plants, and these enzyme systems are ideally adapted to deal with the new threats posed by pesticides.

Resistance in any pest is the same phenomenon as evolution by natural selection, except that pesticides are a human-induced, artificial selective force. Every species is composed of individuals and populations that vary in their characteristics, and the introduction of a new selective agent such as pesticides affects some more than others. Those individuals with a gene or genes imparting some resistance to a pesticide will survive and reproduce at a higher rate than susceptible individuals, producing future generations that are increasingly insecticide-tolerant.

Pesticides are an especially powerful selective agent, since they initially kill most pest individuals. However, the few that survive find themselves growing and reproducing in an environment in which competition for food and other resources has been vastly diminished by the thinning power of pesticides. They also wind up mating with other resistant individuals, which magnifies the genetic resistance of the next generation. Thus, the surviving populations quickly rebound with resistant individuals, leading growers to apply higher levels of pesticides, which in turn select for further resistance. Pesticide use soon becomes a quickening treadmill, in which increasing quantities and frequencies of pesticide application are necessary, and the pesticide industry has to invent new chemicals continuously to keep pace with pesticide-induced evolution of resistant organisms. And because resistance to one pesticide often confers resistance to others, even across different classes of pesticides, frequent changes in chemical applications may not prevent the development of resistance.

Resistance is not a new occurrence among pests, but it certainly is an increasing problem. The first documented example of resistance was in 1908, when the San Jose scale insect was found to have become resistant to lime-sulfur sprays in Washington State. In 1946, house flies in Sweden were discovered to be resistant to the newly introduced DDT; and by 1986, 447 agriculturally important insects and related organisms were known to be resistant to one or more pesticides. Resistance is not confined to insects. Today, at least 100 species of plant pathogens, 55 weed species, two nematode species, and five rodent species are resistant to one or more pesticides formerly used in controlling these pests.

A second, and serious, result of prodigious pesticide applications has  been the reduction or even elimination of predators and parasites that formerly controlled pest populations. These natural control agents include ladybug beetles, ants, hornets, and lacewing larvae, as well as the parasitoid wasps, whose long, sharp, egg-laying tubes pierce the unlucky host's skin and deposit eggs inside its body. Other beneficial insects such as pollinating bees also are susceptible to pesticides, and many farmers today must rent managed bees to pollinate crops because wild bees are almost nonexistent in agricultural areas.

Predators and parasites are especially vulnerable because, first, pesticides can become more concentrated as they move up the food chain. When a pesticide becomes lodged in the tissue of potential prey, the natural enemies that feed on them receive higher dosages than the target pest received. Second, most natural enemies have considerably lower reproductive rates than the pests they live on, so that predator and parasite populations are slow to rebound from pesticide applications, even if resistance should develop. Finally, natural enemies often starve and are virtually eliminated from the environment when their prey or host populations collapse following pesticide exposure. When pests rebound owing to resistance, there are no natural enemies left to keep their populations in check, and it becomes open season on crops or humans for the rebounding and newly resistant pests.

After continued use of pesticides has wiped out a target population, previously innocuous insects may suddenly explode into pest status. For example, the agriculturally insignificant Californian cotton leaf perforator became a serious cotton pest when DDT killed competing cotton feeders but left resistant populations of cotton leaf perforator to grow. Subsequent spray applications of carbaryl eradicated another cotton pest, the pink bollworm, but also eliminated the leaf perforator's natural enemies, forcing farmers to spray even more pesticide to control the burgeoning leaf perforator population.

❧

Many diverse nonchemical, biologically based strategies for pest management have become available in the last hundred years. However, the adoption of alternative techniques in commercial pest management has lagged far behind our academic understanding of their potential. The implementation of alternative concepts, which predominate in the scientific community, is not widespread in solving real world pest situations. The pursuit and adoption of biologically based methods of pest control have been hampered by our commitment to pesticides, our psychological attitudes toward pests, short-term economic concerns, the complexity of nonchemical management techniques, and a lack of comprehension about our impact on the rest of nature.

The issue is not that the use of chemical pesticides is invariably harmful and should be banned. Limited and timely pesticide applications have and will continue to play an important role in protecting our homes, crops, and health. Rather, pesticides are similar to many technological achievements in contributing positively to human life when properly used but inducing damage when overused or used improperly.

The automobile provides a good analogy to pesticide use. Cars have improved our lives in many ways, but misuse of automobiles by speeding or running stop signs or driving under the influence of alcohol kills thousands of people each year. Similarly, pesticides can be beneficial, but spraying pesticides directly on farm workers, storing chemicals where children have access, or using overly toxic substances can be as deadly as a high-speed car crash. A regular car-pool commute to work or an occasional drive across town to see a movie may make our lives easier and more enjoyable, but overusing automobiles for long, daily, single-passenger commutes or frequent drives to the supermarket around the corner are major factors in polluting our air supply. Excessive pesticide use also has had undesirable side effects, such as residues in our food and selection of resistant pest complexes that in turn force us to use increasing quantities of chemicals.

For both cars and pesticides, most of us would agree that their use should be reduced and misuse eliminated. For pesticides, there are a number of interrelated factors that seem to be preventing us from moving to more ecologically balanced modes of pest management. These factors reverberate throughout pest history, but in the last hundred years have matured into critical problems along with our other hits on the environment.

One recurrent and increasingly persistent theme in pest management concerns the impact on nontarget organisms. The classic cases of insects developing pesticide resistance and birds and fish accumulating pesticides and dying at potentially extinctive rates are just two examples of the many instances in which our technological mastery over pests has in the end failed us, as well as the organisms and ecosystems we claim to steward.

Another problematic theme in pest control has been the muddled dialogue between specialists and the public. The general public is easily swayed by well-intentioned but uninformed attacks on legitimate pest management, while specialists often fail to explain adequately to an increasingly skeptical public just what they are doing. In addition, the "experts" frequently base management decisions on information that is less credible and more ambiguous than they believe. This predictably troubled interaction has strongly influenced how we manage pests, often to the detriment of both pest management and the public interest.

A third problem among those who manage pests is their belief in the potential to truly achieve technological mastery over pests through scientific advances, without adverse environmental side effects. We always seem to be on the verge of a new relationship with pests, in which we can target particular pest species without harming nontarget organisms. It is true that our scientific concepts of pest management are evolving toward multidisciplinary techniques that view pests as normal components of ecosystems that can best be controlled through integrated management techniques rather than brute eradication. However, these philosophical advances in pest management have yet to be fully realized in practice, and progress in this area has been slowed by political and economic factors as well as by the inevitable inertia of established methods. Also, scientists have not always been effective at translating potential breakthroughs in basic science into pest management applications that are commercially viable.

Another theme that echoes through the pest management world is that pests are context-specific, and we often define the context much too broadly. We in North America tend to overdefine "pest," so that a single cockroach or a few ants walking across a kitchen counter will elicit a panicked response, and a heavy dose of pesticide. A considerable amount of pesticide use in agriculture is focused on a consumer-driven imperative to produce perfect fruit, with no blemishes and not a single "worm" inside, whereas these are cosmetic issues more than threats to our health or food supply. Much of our pest management is aimed at these types of situations, in which pests are created simply by the unrealistic standards we set. The environmental cost of cosmetic pest management is staggering.

Finally, and in the end perhaps most significantly, our thinking about pests and the practice of pest management has taken place in the context of our broader thinking about the role of humans in nature. Attitudes have ranged from our divine right to manipulate nature for our own ends to an environmental ethic that views humans as caretakers of Eden. Pests are a part of this larger issue, and our decisions about managing them tell us much about ourselves, about how we view nature, and about the problems that develop as we attempt to fit the earth into a mold that is of increasingly human design. We may need to conduct particular battles against pests, but our battles have lately escalated into a costly war on nature itself, and in the end it is a war that we are bound to lose.

CHAPTER TWO

Gypsy Moth

"Would I sit outside and drink my coffee when they were spraying? Yes. I would have my kids beside me. I am absolutely, utterly confident in its safety. It has been tested world-wide. It has been used and used and used. If there was any risk to people I would not be doing this."

Jon Bell, Agriculture Canada (1992)

In the spring of 1992, in the city of Vancouver, British Columbia, 45,550 acres were sprayed by air with three applications of a bacterium called *Bacillus thuringiensis*. The spray used in this program selectively kills only butterfly and moth larvae. The target of this aerial bombardment was the gypsy moth, a pest that already had set numerous legal, scientific, and pest management precedents throughout North America since its introduction in 1869. But the magnitude and controversy of the Vancouver incident stood out in the history of this most experienced of pest species.

The gypsy moth sprays of 1992 generated a world-class controversy in spite of wide consultations, solid evidence of the moth's potential to wreak economic damage, close-to-unanimous scientific agreement from forestry and pest management experts as to the desirability of the *Bacillus thuringiensis* spray program, and innumerable public meetings in which the government sought to explain and justify it. The Vancouver gypsy moth infestation became the center of a battle between an increasingly sophisticated public relations machine run by the government and a media-generated counterattack based largely on undocumented claims by a few environmentalists.

The gypsy moth's assault on Vancouver was a two-pronged offensive, in which two varieties of the species arrived simultaneously from two directions, the European gypsy moth via eastern North America and the Asian gypsy moth directly from Siberia. Neither variety of gypsy moth belonged in Vancouver, or anywhere else in the New World for that matter. The European gypsy moth, the first to arrive, was deliberately imported to North America when cotton became unavailable following the Civil War and the United States needed another source of fiber. The French astronomer and naturalist Leopold Trouvelot, who was living in the textile mill town of Medford, Massachusetts, was conducting various experiments with silkworms, whose cocoons can be unwound and the threads turned into silk. He decided to import egg masses of the European gypsy moth to investigate whether the cocoons of this insect also could be used to manufacture fiber for clothing and other uses.

Neighbors reported later that some of the gypsy moth eggs were blown out an open window by the wind, and Mr. Trouvelot was considerably disturbed by this event. He evidently was aware that the gypsy moth was a serious forest pest in Europe with a voracious appetite, because he immediately reported that these insects had escaped from his custody. Word spread quickly in the entomological community. The renowned C. V. Riley, then State Entomologist in distant Missouri, noted only a year later that the larvae of this moth, "which is a great pest in Europe both to fruit trees and forest trees, was accidentally introduced into New England, where it is spreading with great rapidity."

Riley was a bit premature in his warnings. During the first twenty years of the gypsy moth's history in North America its impact was local. At first, it was a pest only in Trouvelot's neighborhood, and its spread through Medford was noted on almost a block-by-block basis. A neighbor on Myrtle Street reported that "the caterpillars troubled us for six or eight years before they attained their greatest destructiveness. They were all over the outside of the house, as well as the trees. All the foliage was eaten off our trees, the apples being attacked first and the pears next." Another neighbor wrote that the "caterpillars were very troublesome in our yard and in those of our immediate neighbors. At that time they were confined to our part of Myrtle Street, but they soon spread in all directions . . . The caterpillars would get into the house in spite of every precaution, and we would even find them upon the clothing hanging in the closets. We destroyed a great many caterpillars by burning, but their numbers did not seem to be lessened in the least . . . I think that if an organized effort had been made at that time to destroy the caterpillars they might have been stamped out."

In 1889 the moths suddenly reached outbreak proportions within a two-mile radius of Trouvelot's house. E. H. Forbush and C. H. Fernald, two prominent Massachusetts entomologists, reported that this infestation was "almost beyond belief. The 'worms' were so numerous that one could slide on the crushed bodies on the sidewalks, and they crowded each other off the trees and gathered in masses on the ground, fences, and houses, entering windows, destroying flowering plants in the houses, and even appearing in the chambers at night . . . A sickening odor arose from the masses of caterpillars and pupae in the woods and orchards, and a constant shower of excrement fell from the trees . . . The caterpillars devoured the foliage of nearly all species of trees and plants in the worst infested region."

The town residents fought back, often in hand-to-hand combat. Leisure time was spent picking larvae off backyard plantings one by one, with buckets of larvae to show for an evening's work. Some residents banded tree trunks with tarred and inked paper to catch and kill the caterpillars; others fought the insects with fire, water, and coal oil. In spite of their efforts, however, the gypsy moth continued to spread and eat.

The Massachusetts state government entered the fray in 1890. The legislature passed an *Act to Provide against Depredations by the Insect Known as the Gypsy Moth,* and the official battle had been engaged. This act was the first law in the United States to establish the right of regulatory officials to enter private property for the purpose of eradicating a pest, and its echoes were to reverberate as the moth spread throughout the northeastern United States and southeastern Canada, culminating in its arrival at the end of the next century in the city of Vancouver.

The Asian variety of gypsy moth was a more recent and accidental immigrant, but its arrival in 1991 from Siberia also came as a result of political and economic factors. The 1980s in the Soviet Union were a time of turmoil, and the ponderous Soviet agricultural system could not produce nearly enough grain for its inhabitants. The collapsing Soviet Union turned to North America for emergency grain shipments, and the vast quantities required to feed the hungry Soviets could not be handled by existing Pacific ports. The Soviets began using abandoned military port and rail facilities at isolated coastal sites in Siberia, and equipped the ports with bright arc-lites to prevent pilfering at night and to allow the grain ships to be unloaded around the clock. Unfortunately, the new ports were adjacent to forested areas that were hosting a growing outbreak of the Asian gypsy moth.

The gravid female Asian moths were attracted to the bright lights of the ports and deposited their egg masses all over the grain freighters, which then carried the developing eggs to the port of Vancouver. The eggs hatched while the ships were approaching the grain-loading terminals, the young caterpillars were carried by wind into the city, and the Asian phase of the Vancouver infestation was on.

There is nothing about either the Asian or European gypsy moth's appearance that would alert the casual observer to its danger. The adults are a dull brownish-gray or white color, blending in well with the female's preferred egg-laying site, tree bark. The female moth lays flattened masses of eggs in protected crevices, in which 100 to 1,200 eggs spend the winter before hatching in the spring, coincidental with leaf budding. The young larvae are the source of the name gypsy moth, because the hairy, newly emerged larvae climb to the tops of trees, spin down on silken threads, and can then easily be detached and dispersed by the wind up to 16 kilometers.

The larvae spend the spring consuming large quantities of foliage, with each larva eating the equivalent of between 3 and 18 large leaves. When feeding is completed, the satiated larvae pupate and metamorphose into adults. The adult female moths secrete an airborne attractant that lures the males to mate, and the cycle begins again.

The pestiferous nature of the gypsy moth is due to its voracious and somewhat indiscriminate feeding habits. A total of 485 plant species have been recorded as food sources, about 150 of which are woodland tree species that are preferred host plants. Oak leaves are the moths' favorite meal, but they will consume foliage from most deciduous forest trees as well as from orchard and backyard plantings. The European moths eat coniferous foliage only in a pinch, but the Asian moths are quite happy to consume conifers as a regular diet, and so the Asian variety is more of a potential pest to western forests than is the European moth. In addition, the adult female European gypsy moth is almost completely flightless, whereas the Asian variety flies well and is considered to be more dangerous because of its better dispersal capabilities.

The most direct economic threat of both the European and Asian gypsy moth is the defoliation of entire forest regions during an outbreak. Typically, a local population will exist in low numbers for many years, kept naturally under control by parasitic flies and wasps and predacious beetles and mice. This innocuous phase can change into an explosive period of population growth during hot, dry years, especially in forests with a high density of oak trees. Up to 5.7 million larvae per acre may be produced, and virtually all of the foliage in the area may be consumed. After two to three years a combination of starvation and attack by viral and bacterial pathogens leads to the collapse of the population. Outbreaks occur on average every eight to eleven years, but natural variability in population growth and decline make predictions difficult.

The first experiences with gypsy moth outbreaks around Medford, Massachusetts, were almost a template for what occurred in Vancouver a hundred years later. The Massachusetts legislature was stimulated to action by a combination of scientific experts and various interest groups, and the five years following the 1889 infestation were a time of precedent-setting management approaches to the growing gypsy moth problem. But efforts to eradicate this pest failed in spite of a large-scale, expensive, and state-of-the-art campaign.

Appeals to the Massachusetts legislature came from the Massachusetts Horticultural Society, the County Agricultural Society, the entomologist Charles Fernald, who was assigned to the State Board of Agriculture, and the ubiquitous constituent pressures brought to bear on local politicians to "do something." The legislature acted with surprising speed, and on

March 14, 1890, passed legislation providing for a coordinating Commission of up to three "suitable and discreet persons whose duty it shall be to provide and carry into execution all possible and reasonable measures to prevent the spreading and to secure the extermination of the gypsy moth in this Commonwealth." The legislature also devoted an initial grant of $25,000 to run the campaign and provided for penalties to any citizens who attempted to interfere with the work of the Commission. By 1900, $1.2 million had been allocated and spent against the gypsy moth, equivalent to about $27 million in 1996 dollars.

The key elements in defining the campaign were Fernald and a panel of experts from all over the United States that he convened in March 1891 in Boston. The meeting was attended by local entomologists, politicians, and citizens, and outside experts such as C. V. Riley. Riley and Fernald, both of whom had tremendous influence on the other panel members, realized how serious the infestation was and how limited were the chances for successfully eradicating this pest. Nevertheless, they felt that even if they "fail to exterminate it this year, we shall at least diminish its expansive energy." Riley recommended that "recourse must be had to spraying with some of the arsenites in order to bring about the extermination of the moth."

The eventual decision to spray arsenic-based compounds was not made without discussion. Even the polite blue-ribbon panel convened by Fernald had reservations, although their initial deliberations revolved more around whether sprays would be effective than around the health and environmental issues that would preoccupy citizens in the twentieth-century. Panel members debated extensively whether the gypsy moth was indeed a serious enough pest to warrant such attention, what the best techniques would be to control it, and how the public would react to spraying. They also discussed whether vehicular traffic in and out of the area would ultimately defeat any program designed to contain and eradicate the moth by carrying moths out of the target regions and creating new infestations elsewhere. In spite of these reservations, however, the spraying campaign was strongly supported by the experts, and the battle against the European gypsy moth was on.

The 1890–1894 campaign easily equalled or exceeded the efforts in

Vancouver a century later. Part of the decision to spray arsenic-based insecticides was a response to the failure of an initial enormous, well-organized effort to physically eradicate the moths by the simplest of techniques: finding individuals visually and destroying them one by one. For several years, a small army of about 200 people had been sent through the countryside to find and kill egg masses, caterpillars, pupae, and adult moths. Eggs were physically scraped off tree trunks, or creosote or acid was applied to the egg masses. The later life stages were dumped into pails and destroyed.

The extent of this effort and the detailed records kept for each work crew stagger the imagination; 16,638,557 trees were inspected, egg masses were found on 415,724 of them, and a total of 3,833,088 egg masses, caterpillars, pupae, and adult moths were destroyed. In addition, roadblocks were set up to find and prevent gypsy moth egg masses from being carried out of the region by vehicles, and countless trees were banded to catch caterpillars migrating up and down tree trunks. Finally, large areas of forest and scrub brush were cut and burned to destroy heavy infestations.

It quickly became obvious that this approach was not working, and Riley's recommendation to spray arsenites became the focus of the campaign. The entomological experts began to believe in arsenic, and stated in an early report that extermination with sprays "was really possible, provided the work was continued for several years with sufficient appropriations to keep the entire territory under careful supervision." Not everyone agreed with this assessment, however. The first sprays were conducted using Paris Green (a formulation of copper and arsenic), but local citizens complained that it was not effective. A mass meeting of spray opponents was held in Medford to protest these sprays, and one citizen was arrested and fined after attempting to cut the hose attached to one of the spraying tanks and threatening violence to the spray personnel who had entered his land. Others tried to neutralize the sprays by turning their garden hoses on the sprayed trees and shrubs to wash off the solution.

Fernald's meeting of experts spent some time discussing the public response to the failed spray program and how to deal with it. Riley's suggestion was to agree with the public comments, but he thought this opposition could be silenced once people were "given explanations why they had that experience. It was simply due to the impurity of the Paris Green and the imperfect manner of applying it. You will always have more or less failure until you put this matter into the hands of men who can give their whole time to it." The government Commission then responded to the outcry by producing bulletins and distributing handouts containing information about the sprays, especially quotes from the experts assuring the public "as to the lack of danger to man or beast attending the use of Paris Green."

Whatever the real or imagined dangers of Paris Green, it did not prove to be effective, and it was abandoned after 1893 in favor of a new insecticide developed specifically for use against the gypsy moth, arsenate of lead. The advantages of this compound were that it could be used at any strength without burning foliage, it strongly adhered to leaves without being washed off, even in rainstorms, and it would remain on foliage for the entire season. Considerable ground-breaking experimentation was done with lead arsenate to determine the best dose, spraying methods, formulations, and carriers, but it soon became apparent that it too was failing at the task of killing all the gypsy moths.

In the midst of these discussions there was at least some public perception that arsenic-based compounds were dangerous. Forbush and Fernald reported in their 1896 book *The Gypsy Moth* that "prejudice against spraying in Medford was intensified by the belief that there was danger of fatal poisoning to man and animals." However, they treated these as "sensational reports. Statements were made in the daily press that a man had died from the effects of chewing leaves taken from trees sprayed in Medford, and that a child had been fatally poisoned by eating bread and butter on which some of the spray had fallen from the trees. On this at least one newspaper editor advised his readers to shoot on sight the workmen employed in spraying."

Forbush and Fernald dismissed these claims, but also admitted that "there are other dangers arising from the widespread and careless use of arsenical insecticides which have been almost entirely ignored. Entomologists and pomologists officially connected with the experiment stations of the country, the agricultural press, and writers of works on pomology and horticulture all join in recommending some of the most deadly poisons as insecticides, but they add scarcely a word of caution in regard to their use."

Lead arsenate continued to be the insecticide of choice in repeated campaigns as the moth spread, and seemed to have some success. By 1900 little defoliation could be found in the infested areas, and the Massachusetts legislature, believing the moth to be defeated, terminated the program. Unfortunately, the moth populations were only in their latent phase, and 1900–1905 saw another population explosion in Massachusetts. The moth was spreading as well, arriving in Rhode Island (1901), New Hampshire (1905), Connecticut (1906), and Vermont (1912). The moth continued to make political history; on August 20, 1912, the U.S. Congress enacted the Plant Quarantine Law, which was designed to prevent the movement of insect pests from infested to noninfested areas. This law is still in effect today and has been credited with reducing the accidental transport of gypsy moths and other pest insects.

Plant quarantine was not enough to stop the inevitable migration of the gypsy moth, however, nor were the two barrier zones that were implemented following joint meetings of personnel from federal and state Departments of Agriculture and their Canadian equivalents. The first barrier was set up in 1923, in a line running from Canada to Long Island along the Hudson River and Champlain Valleys. The second zone was implemented in 1932 in western New York and eastern Pennsylvania. Professionals in both programs liberated millions of imported parasites and predators, sprayed infested areas in the zones with lead arsenate, and inspected vehicles and household goods leaving the quarantine zones, but the moth easily jumped beyond the barrier zones.

As the century progressed, DDT took the place of lead arsenate, followed by the insect growth regulator Dimilin and then Carbaryl, a slightly more environmentally friendly but still broad-spectrum insecticide. In spite of heavy treatments, the moth continued to spread at an average rate of about 10 kilometers per year. Today all of the northeastern United States and southeastern Canada is infested, and the moth continues to spread slowly from that region. Spot infestations appear far from the infestation center, arising from egg masses transported accidentally throughout the United States and Canada.

The first European gypsy moths to be found in Vancouver came in the form of eggs that appeared on a canoe in 1978. This stimulated a highly controversial but limited spray program that seemed to eliminate the moths at that site. But other moths began appearing in subsequent years—transported by vehicular and equipment movements from the east, especially by military personnel and vacationers to eastern Canada returning to the west. A large-scale monitoring program was instituted using traps baited with synthetic female sex attractant, and between 0 and 166 male European gypsy moths have been trapped each year since 1978 around the Vancouver area, with small-scale eradication programs conducted where populations were becoming established.

The massive urban spray program of 1992 was inspired not by European moths but by their Asian cousins. The very first Asian moths appeared in Vancouver in 1911, as eight egg masses intercepted on imported Thuja trees originating from Japan. These were destroyed without causing an infestation. A few other egg masses were found on ships in 1982 and 1989, but these too were intercepted before hatching. In the spring of 1991, however, over 2,000 egg masses were found on ships entering the port of Vancouver from the Soviet Union, and thousands of larvae were seen ballooning toward shore.

By the time the Asian gypsy moth hit Vancouver, it and its European cousin had developed a reputation as a serious pest, based on estimates of the damage that the European gypsy moth had caused in eastern North America as well as anecdotal reports from Asia. The major impact of gypsy moths is complete and repeated defoliation of trees during outbreaks. While most trees can withstand partial defoliation, or even a single year of total defoliation, leaf loss in consecutive years weakens trees and makes them susceptible to lethal attack by other insects and pathogens, especially bark borers and root fungi. Even surviving trees are affected, sustaining a 30–60 percent reduction in growth during outbreaks, which postpones harvesting and reduces wood quality. Indirect losses accrue from the increased fire hazard in defoliated forests, damage to the nesting habitat of birds and other wildlife, reduced food sources for these animals, increased erosion and loss of water-retention in the soil, and damage to the aesthetic and recreational value of woodlands, which has an adverse impact on tourism.

In the United States, 59 million acres of trees were defoliated between 1924 and 1990. In 1980 alone, 25.9 million acres, or 40,500 square miles, were defoliated in New England and Michigan. In Ontario, 770,000 acres were defoliated in a 1991 outbreak, and the Ontario government had spent almost $2 million to spray only a tenth of the outbreak area. In Virginia, estimated losses due to gypsy moths in only seven counties were $14 million, with 25–35 percent of the trees dying following defoliation. An outbreak in the Newark, New Jersey, watershed killed over 1 million oak trees, and a nearby Pennsylvania outbreak resulted in an 83 percent reduction of sawtimber and pulpwood value due to dead or declining trees.

The sheer magnitude of the defoliation in these outbreaks was a major factor in the strong response of western Canadian regulatory officials following the gypsy moth finds in and around Vancouver. But trade considerations, rather than forest health, was the main reason for their concern. In the United States, which is Canada's main trading partner, quarantine regulations prevent lumber and other goods from being imported unless they can be certified as free of gypsy moths. Loss of trade with the United States was, and remains, the most potentially damaging aspect of an incipient gypsy moth infestation in western Canada.

Although spraying was clearly the most obvious response to Vancouver's gypsy moth infestation, regulators in Agriculture Canada—the official government wing charged with detecting and eradicating new pests—had learned over the years that wide consultation is advisable before any type of spray program. Consequently, Agriculture Canada consulted with the British Columbia Gypsy Moth Committee, a group very similar in composition and function to Fernald's advisory panel in the nineteenth-century Massachusetts infestation. The Gypsy Moth Committee consisted of federal, provincial, academic, and industry participants, but it also officially or unofficially consulted local environmental groups prior to making recommendations. The B.C. Ministry of Forests also became involved in the gypsy moth program, because of their obvious interest in eradicating gypsy moths as well as their expertise in conducting large-scale spray programs.

Agriculture Canada, then and now, makes a strong distinction between programs to prevent establishment of a pest and programs designed to control one that is already established. In the case of the gypsy moth, this is not just a fine semantic point, since Canada's trading partners, including the United States, will not accept shipments of lumber and other goods if the gypsy moth is considered to have become a resident insect of British Columbia. Thus, as long as Agriculture Canada is conducting an eradication rather than a control program, they can still certify that British Columbia is free of gypsy moths, and this certificate accompanying shipments satisfies Canada's export markets.

Not surprisingly, the first level of controversy in Vancouver's gypsy moth saga developed over whether eradication was possible. Most pest managers supported attempts to eradicate the gypsy moth, arguing that it had not yet became established in British Columbia. But spray opponents argued that gypsy moth populations in the Vancouver area had already become established and that continued spraying against the moth was futile and unwarranted; the gypsy moth should be left to follow its own intrinsic cycles of arrival, outbreak, and decline. These dissenters argued that further efforts to eradicate the pest were just a charade to satisfy trade regulations artificially imposed by the United States. A second issue centered on an ongoing debate in the management of any insect: Is it more biologically and economically effective to allow population outbreaks to expand and collapse on their own cycles rather than to intervene with costly spray programs of questionable efficacy? Agriculture Canada thought that it was.

Whatever the merits of the "spray and pray" versus the "let it be" schools of pest management, the eradication approach dominated among the vast majority of scientists and regulatory personnel following the 1991 moth finds in Vancouver. However, the regulators had become extraordinarily sensitive to public opinion about sprays since the first gypsy moths appeared in the late 1970s. Strong and effective lobbying from environmental groups at that time had persuaded Agriculture Canada to avoid spraying the "hard" pesticides that had been initially proposed, and to use instead a relatively new and environmentally more benign substance, *Bacillus thuringiensis.* Indeed, environmental groups such as Greenpeace had offered to buy *B.t.* for these early spray programs, so it was surprising when the proposal to spray Vancouver with *B.t.* in 1992 met fierce opposition.

Mobilized initially by Rachel Carson's *Silent Spring*, environmentalists from the 1960s into the 1990s had condemned the wanton use of poorly tested, broad-spectrum, medically and environmentally dangerous substances to control pests. Their concerns had stimulated scientists to search for alternatives, and *Bacillus thuringiensis* was one of the results. It was the most recognized, heavily tested, successful, and safe alternative to synthetic chemical pesticides available, yet in Vancouver, *B.t.* came under public fire.

Bacillus thuringiensis is a common spore-forming bacterium that is nonpathogenic to warm-blooded animals but is highly pathogenic and specific to certain larval butterflies and moths. The bacterium has many varieties and subspecies, but the variety used against the gypsy moth was first isolated in 1962 from a diseased laboratory colony of pink bollworm larvae. Thirteen companies in the United States and Canada produce various *B.t.* products, with 17 different formulations registered for use in Canada to control lepidopterous forestry pests. *B.t.* is produced commercially in large fermentation vats, where it is allowed to sporulate. The concentrated spores are formulated into either aqueous suspensions or oil emulsions that can then be sprayed by ground or air onto foliage. They enter the insects when the larvae feed on *B.t.*-coated leaves.

The effectiveness of *B.t.* against gypsy moth is due to a protein secreted by the bacterium. This protein, when consumed by a larva, binds to the membranes of the insect gut and destroys gut cells, thereby preventing further feeding and eventually killing the insect. The spores are not toxic unless their proteins are dissolved under the high pH conditions (pH 9.0–12) characteristic of butterfly and moth digestive systems. Thus, *B.t.* is both effective and highly selective, and has no known effects on nontarget organisms except for some other moths or butterflies, even at high concentrations. It has been extensively used world-wide for over 35 years in contexts ranging from backyard gardens to organic farms to large-scale aerial applications in forestry, with no evidence of any negative medical impact and only slight environmental effects due to short-term reductions of other moths and butterflies. Moreover, it has undergone every conceivable test for toxic, carcinogenic, or mutagenic effects on a wide variety of nontarget organisms, with no adverse effects reported.

Nevertheless, the very idea of being sprayed with bacteria from the air generated an enormous firestorm of public outrage in Vancouver, in spite of repeated and intensive assurances from medical and insect control personnel that the sprays were safe. The highly emotional reactions to the spray program were based on two general concerns. First, the public basically distrusted scientists and regulators when it came to insecticides and other issues. Second, the public feared medical effects following human contact with bacteria.

The public's distrust of insecticidal sprays may not have been warranted in this situation, but certainly the history of gypsy moth spray programs would not inspire confidence that medical and environmental issues were adequately considered by control personnel. Gypsy moth control programs read like a litany of insecticidal disasters, from the first copper and lead arsenate compounds, through subsequent programs using the infamous DDT, until the more contemporary sprays with Dimilin, an insecticide now highly restricted because it was found to be potentially carcinogenic. Indeed, the limited 1979 Vancouver gypsy moth ground spray program initially proposed the use of Dimilin, and the historical memory of Vancouverites contained considerable anxiety about Agriculture Canada's disregard for public safety. The 1979 headlines such as "Kitsilano Spray Genetic Time Bomb," "Cancer Causer?" and "Dread Dimilin Spray" were factual and not easily forgotten by the public.

Public concerns about the medical effects of *B.t.* were not realistic, however. The main concerns about *B.t.* focused on possible infections, especially in immunosuppressed individuals such as those carrying the HIV virus. However, a thorough review by Dr. R. G. Mathias (Chair of the University of British Columbia's Division of Public Health Practice) of 35 years of medical reports and about 50 health-related studies throughout the world revealed no substantive risk from *B.t.* or the carrying materials in which it was formulated. In a letter to Agriculture

Canada reporting his findings, he wrote: "In studies which have been done with individuals working with this organism, even under conditions of heavy exposures, *B.t.* has not been an infecting organism. It is one of many soil organisms that do not colonize or infect humans. *B.t.* is adapted to insects, not to humans."

Another review by the Provincial Health Officer H. M. Richards came to the same conclusion, and pointed out that "the scientific evidence linking gypsy moth to human illness is stronger than that for *B.t.*" Gypsy moths themselves can cause skin rashes from excessive contact with larval hairs; while discomfiting and a nuisance, these rashes are not life-threatening. Richards went on to point out that the main danger to public health from gypsy moth control programs is not *B.t.* or gypsy moths but accidents involving the aircraft and motor vehicles used to deliver the spray. One helicopter crash had occurred in the Vancouver area during a previous chemical spray program.

The final decision to spray with *B.t.* was made following intensive professional meetings and public input. Agriculture Canada and the B.C. Ministry of Forests had learned from previous spray programs that considerable effort needed to go into pre-spray consultations and information campaigns in order to respond to the expected opposition from organized environmental groups and to the questions and concerns of the broader public. Most of the established environmental groups chose to remain on the sidelines of the *B.t.* debate. Groups such as Greenpeace, the Western Canada Wilderness Committee, and the Sierra Club were invited early on to meetings explaining the rationale and methods of the proposed spray program. They all elected to remain silent, with their silence interpreted as either quiet support for the program or at least lack of major concerns about it.

However, three new, small, and environmentally militant groups took the issue on. The Society Promoting Environmental Conservation (SPEC), Citizens Against Aerial Spraying (CAAS), and the Society Targeting Overuse of Pesticides (STOP) managed to catch the attention of local, national, and even international media in a series of well-crafted news releases and letters to local newspapers.

The claims coming from these groups included attacks on the scientific rationale for the spray program. For example, one letter from SPEC member Dermot Foley to the *Vancouver Sun* pointed to a 1974 review of gypsy moth ecology in the *Annual Review of Entomology* that "contained much information about natural predators of gypsy moth found in North America." His letter implied that these predators would prevent moth outbreaks in British Columbia, although they had not done so elsewhere in North America. He also quoted a retired forest entomologist, Kenneth Graham, as saying that supporters of the spray program had "insufficient appreciation of the history, ecology, or population dynamics of forest entomology."

The central focus of the environmentalists' campaign, however, was potential health effects of spraying *B.t.* in urban areas, and here they were on even shakier ground. In the same letter, Foley highlighted one report from the *American Journal of Ophthalmology* that found *B.t.* in an eye lesion, although that article and subsequent interpretations of it stated that the *B.t.* infection was likely secondary and not the cause of the lesion. Similarly, he pointed to another study that found *B.t.* to be related to a bacterium that is pathogenic to humans, although there is in fact no reason to believe that *B.t.* itself is harmful or has even a slim possibility of evolving into an organism that is harmful to humans. The group Citizens Against Aerial Spraying went further and sent a FAX to the United Nations calling the aerial spray program a "human crop-dusting experiment," and urging the U.N. to intervene.

Media from all around the world picked up the stories as the April spray dates approached, and highlighted the gypsy moth spray program in news magazine stories and lead reports on radio and television news programs. At one point calls from the media were coming in to Agriculture Canada once every six minutes, from as far away as Japan. The government saw what they viewed as their rational, carefully justified program dissolving in the face of intense media pressure generated by perhaps fewer than ten individuals. The time had come to respond.

The government turned to a sophisticated former journalist and media maven, Nancy Argyle, who works for the B.C. Ministry of Forests as a communications expert. Argyle is articulate and confident, with the empathic air of a counseling psychologist. She is an outspoken proponent of communications as a discipline that differs from public relations in providing information about issues rather than hype and blunt advocacy.

She dresses in a Ministry of Forests uniform reminiscent of Smoky the Bear, and indeed it was the skills she developed in crisis management during forest fires that proved to be just what the gypsy moth program needed to overcome the growing protests against the sprays.

She did not have an easy job. "I would say it was the spray project of the decade, maybe the century. I used to sit at my desk and envision that we were going to take these huge aircraft and fly 300 feet over a major metropolitan city and spray them, and I just couldn't visualize that taking place." Her first task was to persuade the scientists involved in the program that they had a media battle on their hands and that they needed to be trained as communicators. "Some of the worst battles I fought were all internal, trying to convince our main team, which is basically 25 scientists, who all think they're very good communicators, that they're not. Here we had 25 very skilled professionals in their field, a lot of expertise, and they just felt that if they explained everything really well everything would be okay. One scientist said how he had been trained to respond to any kind of question or concern by burying the pertinent information, and in fact, that's not what you do."

What they did instead was take a warmer, more personal approach rather than the overly detailed, analytical tack that had turned the public off in previous spray programs. "I think the most important point in our approach to communications was the fact that we acknowledge people's emotions. That was critical and for a lot of technical people was a very difficult approach. They're used to dealing with hard facts and their training was all geared towards that . . . Whether you believe that *B.t.* is safe or not is really irrelevant. The person on the other end of the phone doesn't think it is."

Argyle's approach was not cheap. Over $250,000 was spent to justify and explain the 1992 spray program to the public, but the campaign worked. They began by distributing 250,000 copies of a *Gypsy Moth Update* bulletin to every household and business in the spray zone, with two subsequent bulletins distributed prior to the spray. In addition, a gypsy moth phone hotline was established (666-MOTH), staffed by individuals trained in personal crisis management, which ultimately took over 26,000 calls from the public. They also held innumerable open public meetings and meetings with focused interest groups such as garden clubs, tourism groups, nursery associations, and the like. Finally, and perhaps most important, they developed a policy of complete openness and availability to the media. Not only was the communications team trained in how to respond to media and public queries, but they were aggressive at providing stories that made the spray program appear essential and safe.

In the end, the gypsy moth communications team succeeded in calming the public's fears sufficiently so that the spray program could proceed. "It was a very invasive program, considered by many people a violation of their rights. Here was this huge aircraft coming over at a very low altitude and spraying them, and spraying their cars and their picnic bench and the cat, and, you know, the kiddies in the pool. That's enough to tick off most people . . . I have to say that the majority of Vancouver residents were not so much supportive of the program as they just reluctantly accepted it. I think that's the best you can hope for in that kind of project."

The aerial sprays began on April 18, following days of rain delay, and continued into May, with three applications of *B.t.* applied to much of Vancouver by helicopter and DC-6 aircraft. In the middle of the spray program, a CAAS press release stated that "the aerial spray may have claimed its first known casualty," a boy who had died at Children's Hospital on May 2. CAAS went on to state that the boy's "mother had taken him for a walk in the recently sprayed area where he lived. Afterwards, the child was bathed in tap water, which has now been contaminated by repeated sprayings and contains *B.t.* spores . . . The responsibility for the death of this child could rest with Agriculture Canada and the premier of the province." In fact, the boy had leukemia and had recently undergone a bone marrow transplant which required drugs to suppress his immune system; the child had died of a bacterial infection unrelated to *B.t.*

In the end, the 1992 gypsy moth spray program cost over $6 million, but it appeared to be successful. No gypsy moths were found in the city of Vancouver in 1993, and there was no evidence of any health or environmental damage from the *B.t.* sprays. A few moths were found in the Vancouver area in 1994 and again in 1996, and both European and Asian moths have been found in nearby municipalities and across the Strait of Georgia on Vancouver Island. These finds are considered by Agriculture Canada to be independent importations of moths into the area, rather than survivors of earlier spray programs.

Spraying is just one part of the massive effort mounted each year by Agriculture Canada to maintain the perception that the gypsy moth is not established in British Columbia. The first line of defense is prevention. Physical inspection of incoming ships from Asia has been standard practice for many years, to prevent the arrival of more Asian gypsy moths. Shipments of household goods and military shipments from eastern Canada are frequently inspected for egg masses of the European moth, since 80 to 90 percent of new infestations are thought to arrive via this route. Agriculture Canada has proposed legislation that would make such inspections mandatory, and wants to use a postal code system to determine where new immigrants to British Columbia have come from. If their previous postal codes reveal that they originated in gypsy moth outbreak areas, then inspectors would be sent out to examine their belongings and to set out traps to monitor for gypsy moths around their property.

Detection of moths is the next line of defense. Agriculture Canada currently maintains monitoring traps at a density of approximately one trap per square mile around inhabited areas of British Columbia, including campgrounds. The open-ended, waxed cardboard traps are about the size of a small milk carton and are baited with a lure containing synthetic attractants, mimicking those given off by stationary female moths to attract the flying male moths to mate. A male attracted to a trap gets snared by the sticky lining inside and, instead of mating with a female moth, becomes a statistic in the gypsy moth detection program.

Finding a male moth does not necessarily initiate an immediate spray program, however. Rather, the regulators increase the trap density in that area the next year to up to 64 traps per square mile, and further finds may lead to a spray decision. Each year the equivalent of six person-years is devoted to this detection program, with over 23,000 traps set out in 1992 alone following the *B.t.* program.

Localized sprayings of about 1,700 acres in total were needed in 1994 at five sites in southwestern British Columbia, including 34 acres in south Vancouver. These *B.t.* applications continued to generate controversy, with petitions to halt each spray routinely submitted to the B.C. Environmental Appeal Board by SPEC, STOP, and a few local residents. Each appeal was denied, and the sprays proceeded.

The campaign against gypsy moths in British Columbia, although not a total victory, seems to have fought the moths to a standstill, at least for now. The pattern of new moth finds, local spraying, and ritualized public objections will likely continue for some time, until the moth eventually evades our detection and eradication programs and officially becomes an established resident in British Columbia. The cost of these eradication programs is and will continue to be substantial, but the eradication costs to date have been much lower than the predicted economic impact of an established population.

A 1994 report commissioned by the B.C. Ministry of Forests suggested that although the European gypsy moth will probably never achieve outbreak proportions in B.C.'s coniferous forests, trade sanctions imposed by the United States would have an enormous adverse economic impact if it were to become established there even at low levels. Much of B.C.'s $2.6 billion softwood lumber industry, its $85 million nursery market, and its $4 million in Christmas tree sales in the western states could be at risk. The cost of maintaining these markets through significant investments in quarantine, inspection, certification, monitoring, and suppression programs likely would exceed the current prevention, detection, and eradication costs by a factor of at least ten to one. The threat of trade barriers presents a significant incentive for British Columbia to continue its policy of eradicating the gypsy moth.

The nineteenth-century entomologists Fernald, Forbush, and Riley might have succeeded in eradicating gypsy moths in Massachusetts if they had had the same tools and knowledge used in Vancouver a century later, or at least the minimal environmental impact of *B.t.* would have been preferable to the chemicals they employed. In that sense, we seem to have made some progress in pest control during the last century. Nevertheless, even the most optimistic of regulators realizes that the moths will inevitably become established in western Canada. The rate of annual moth finds is increasing, both in numbers of moths caught and in the geographic range in which the moths have been discovered.

Eventually, our current eradication perspective will shift to a control mode, and we will accept the presence of gypsy moths, just as residents of the eastern United States and Canada have learned to accept the depredations caused by this voracious leaf feeder.

Our future relationship with gypsy moths will continue in a state of flux, with eradication mentality gradually yielding to control. Serious outbreaks will be sprayed by whatever chemical or biological control method is most biologically sound and cost-effective, with the least environmental damage. New methodology may improve our success, and perhaps at some distant date we will be able to effectively manage and possibly even eradicate this well-adapted insect. More likely, we will continue to co-exist with gypsy moths, not due to some sense of shared rights to our forests and backyards but because we have not been able to develop the magic bullet to rid ourselves of this pest forever. Dermot Foley of SPEC said it well: "It's going to be a war against nature that will go on forever."

CHAPTER THREE

Relatively Harmless Creatures

"To some degree phobias are responsible for some of the pest control operator's revenue. A considerable portion of the industry's income is derived from relatively harmless creatures."

Harry Katz, *Handbook of Pest Control* (1990)

The Linindoll Pest Control Company of Schenectady, New York, had the dubious honor of treating the world's largest recorded cockroach infestation in 1979. The Schenectady Board of Health had been concerned about a large infestation of German cockroaches in a two-family dwelling, and hired Linindoll to deal with the problem. The house was inhabited by 24 dogs, 20 cats, and, as Harry Linindoll delicately put it, "two reclusive elderly ladies, covered with sores from flea bites, who probably were a little unbalanced in regard to sanitary conditions."

Also in the house were approximately three million cockroaches that dwelled in the walls, floors, ceilings, attic, basement, and just about everywhere else. Many of the cockroach residents decided one hot summer evening that it was time to move, and overflowed to nearby yards, houses, and even sewer lines. The unhappy neighbors complained to officials, and Linindoll was called in at 2:30 one morning by a sympathetic policeman who doubled as health officer on the graveyard shift.

It was immediately apparent to Linindoll that the usual cockroach control technique of localized insecticidal applications was a bit low-key for this infestation. Not only did the house itself need some serious treatment, but adjoining houses, trees, lawns, and even the ambulance that took the ladies away needed to be scoured of roaches. The treatment used in this situation had to be innovative and massive, but Linindoll was up to the job. The home was bulldozed and destroyed while the pest controllers sprayed a rain of the chemical pesticide Diazinon onto the fleeing cockroaches.

The Schenectady cockroach bombing was, of course, an extreme situation, but cockroaches and other home pests seem to elicit extreme reactions. Our society's ethics about conserving nature do not extend into the home, where we encounter the pestiferous part of nature firsthand. Urban dwellers have led many of the crusades against using pesticides in agriculture and forestry, and yet, ironically, the urbanite turns almost immediately to chemicals when faced with a few cockroaches in the kitchen, some fleas in the carpet, or occasional insects in the lawn or garden.

As we saw in Vancouver in 1992, a gypsy moth spray program within city limits or even a weed-killing herbicide treatment at a local park can generate enormous public reaction, rapidly escalating into panic about the use of pesticides in our civic areas. Yet, this same concern for the health of our environment, our children, our pets, and nature itself does not carry over into the domestic arena. At home, chemicals rule, and minor pest problems can induce a chemical response that would create a media sensation if it were conducted on public land.

Take the lowly cockroach. This insect alone is responsible for a $1.5 billion industry in the United States and provides one-third of the annual income for 15,000 U.S. pest control companies. In 1990, one-quarter of all residences in the United States reported treating a cockroach problem, and the same number of homes were treated for fleas. Sixteen million households hired a professional pest control company that year to control cockroaches, ants, or fleas, mostly using chemicals. In Canada, retail sales of insecticides to individuals average about $45 million per year. In 1991, 202,000 pounds of active pesticide ingredients were sold in the province of British Columbia for domestic use, about one tenth of a pound for every person in the province, enough to kill approximately 18.4 quadrillion cockroaches.

Overuse of chemicals to control pests in and around buildings occupied by humans is not initiated by an evil chemical industry–exterminator cartel but rather is driven by consumer demand. Those involved in structural pest control, as this industry is called, are striving to use management techniques that would limit pesticide use. This is not the image the public carries of structural pest controllers, however. Bob Berns, a staff writer for the industry trade journal *Pest Control Technology,* described the public image of pest controllers as "Svengalis intent on doing unappealing jobs using malodorous potions and mysterious, quasi-scientific techniques." The term "spray jockey" comes up repeatedly in conversations with and about exterminators. The owner of one pest control company described his relationship with his customers by saying, "People are embarrassed that they have insect problems. What you're really offering is a service to the squeamish." Many structural pest management companies still use unmarked vehicles on jobs to minimize embarrassment to customers.

While society views pest controllers as unsavory exterminators, those working in this industry see themselves differently—as protectors of human health, food, and structures. Robert Snetsinger, in a history of pest control, *The Ratcatcher's Child,* defines the industry as "the art, science, technology and business that protects the health and comfort of mankind, and preserves his property from harm and destruction by insects, rodents, birds, weeds, wood-destroying fungi and related pests. The industry is service oriented and, to a large measure, acts as an attendant or therapist to the urban environment." Berns goes even further: "After years of groping for a clear identity and dodging a negative image, the industry appears, at last, to have come together with something concrete. We are now Guardians of the Environment."

Exterminators as guardians and therapists of the urban environment? These novel if somewhat romantic sentiments reflect a real evolution of the pest control industry from its earliest roots. Historically, ridding humans of pests was more of a life or death decision rather than a decision about which of several pest-management techniques to use. As Snetsinger put it, "In these days of refrigeration, packaged foods, disinfectants, pesticides, municipal garbage and sewage services, washing machines and other technology which provide high levels of sanitation, it is difficult for those living in developed countries to envision a time only a few generations ago when mankind lived in filth and squalor . . . The development of a professional pest control industry was an important element in mankind's social development and is associated with an increasing concern for a better life on this earth and the ability to make improvements in human living conditions."

A litany of pests with disastrous effects on human health and habitat reads like a "Who's Who" from some museum of pest horrors. The plague or black death, for example, which was carried by fleas that normally live on rats but will switch to human hosts, killed 25–30 million people in Europe in the 1300s, about 25 percent of Europe's population at that time. Rats and mice regularly consumed up to half of ancient Egypt's stored grain. Malaria, carried by mosquitoes, continues to kill millions of people a year. And even today, with extensive chemical control available, termites are responsible for about $1 billion in damage to buildings in the United States alone.

What is different about the modern structural pest industry compared with historical pest control is that our society now has to struggle with the balance between killing pests and the damage that heavy-handed chemical pest control can cause. Today we are winning the historical war against pests, because of improved sanitation and the discovery of a vast array of chemical weapons. We now have the luxury of making decisions about how we want to manage pests, choices that need to balance pest destruction with environmental health.

The modern structural pest control industry has two streams, the chemical and a more integrated approach, with extremes on both ends. Representing the extreme chemical end is Harold Stein, past president of the National Pest Control Association, who wrote, "It may seem hard to believe now, but in 1969 NPCA leadership was actually excited about the formation of the Environmental Protection Agency. Sad to say, five years later it became necessary for a small contingent of NPCA officers and staff to meet in the White House and seek Presidential relief from a bizarre assault by this same agency upon not only the professional integrity of the pest control industry but the personal and constitutional rights of our members . . . The truth is that we all recognize how the public today is suffering from the hysteria and hyperbole of several decades of eco-terrorism."

At the other extreme is David Buchanan, from All-Natural Pest Control in Vancouver, who advocates the use of chemicals only if they are "natural," which he defines as being on earth for a minimum of 15 million years. His company's pest management philosophy is to use only chemicals or methods that are "suitable in the long run to be carried on a bicycle in a third world country, and couldn't kill fish. Petrochemicals are not meant to be put on the face of the earth, and petrochemical products simply should not exist."

Midway between these two extremes are the environmentally friendly techniques of Integrated Pest Management (IPM), which have been used in agriculture and forestry for many decades. IPM is beginning to be adopted by structural pest managers, and interestingly the impetus to do so is coming from the industry itself.

❧

Cockroaches provide an excellent example of what IPM can do, and also of today's struggle to bring new and safer techniques to domestic pest control. Contemporary cockroaches are primitive insects, relatively unchanged in their characteristics from the earliest cockroaches that appeared 350 million years ago. They are the close kin of other ancient insects such as grasshoppers, mantids, crickets, and termites. The roaches we see today are almost identical to those that walked the earth long before the dinosaurs, and there is every reason to believe that cockroaches will continue to exist on earth in more or less their present form long after humans have become extinct.

There are about 4,000 species of cockroaches in the world, but four of these species make up the bulk of pest control problems in North America: the German, brown-banded, Oriental, and American cockroaches. All four species originated elsewhere, in wet tropical habitats of Africa or Asia, and likely were carried to North America on ships during the last few centuries. Each species is unique in what it eats, where it lives, and how pest controllers deal with it, but to the nonexterminator all cockroaches basically have the same attributes that have contributed to their enormous success and have made them difficult to control. Their biology is elegant in its simplicity and is ideally adapted for life as a human pest. Cockroaches are flat, long-legged, and greasy; they flee from light, prefer moist environments, protect their young, and eat almost anything.

Being flat, greasy, and fleeing from light may be disgusting to us, but these traits are very important defense mechanisms to a cockroach living in a tropical forest or a kitchen. Cockroaches are quick to flee from attack, using their long legs to run rapidly from danger. They retreat and hide in thin cracks when threatened, using their oily external coating to grease their slide into the narrowest of crevices. Cockroaches prefer warm, moist habitats reminiscent of their ancestral home in the tropics. To us, a building in New York City is not at all similar to a rain forest in New Guinea; but to a cockroach, hot water pipes, moist sinks, drains, stoves, refrigerators, and shower stalls provide an ideal habitat.

Furthermore, the scavenging lifestyle of cockroaches, who are highly omnivorous, is well-suited to our throw-away society. They prefer cereals and sweetened products, or meat, but will feed on such diverse morsels as cheese, beer, leather, wallpaper, documents, postage stamps, and book bindings. Cockroaches even eat hair, sometimes still attached to its owner; they have been reported to regularly consume the eyelashes and toenails of sleeping children in Brazil and elsewhere. Their feeding habits lead them to live in and near food-preparation and storage areas, especially home kitchens, hotels, restaurants, supermarkets, bakeries, cruise ships, and food warehouses.

These lowly insects are remarkably sociable considering their primitive evolutionary status. The males of some species have glands under their wings that provide food to the females. These glands are exposed to females during mating, and a mating female will feed from these glands while the male copulates with her. Once mated, the female's eggs develop in a hardened case that looks somewhat like a grain of rice. The female carries it around with her and protects it until the 10–40 young inside are ready to emerge as nymphs. The nymphs take one to fifteen months to develop into adults, depending on the species, and then the cycle starts anew. Cockroaches are gregarious by temperament and usually are found aggregated, interacting in whatever ways cockroaches deal with their own kind. They find one another by attraction to the aggregation pheromones cockroaches produce.

Considering the vast amount of money and chemicals that are expended in cockroach control, and the extreme revulsion people exhibit in the presence of even a single intruder, it is surprising that cockroaches do not pose a major health threat to humans, except perhaps in heavy infestations. Cockroach specialists cite innumerable pathogenic organisms that roaches are "capable of" transmitting or have the "potential to" transmit, including the causative organisms of gas gangrene, diarrhea, pneumonia, leprosy, typhoid, roundworm, hookworm, food poisoning, and paralytic polio. But food poisoning caused by *Salmonella* bacteria is the most likely health problem that cockroaches could cause, by feeding on products infected with bacteria and then depositing bacteria-laden feces elsewhere. This mode of transmission could be particularly problematic in restaurants and other food-preparation facilities, or in poultry barns.

The medical community, however, is not overly concerned about cockroaches transmitting diseases to people. Many of the pathogenic organisms potentially carried by cockroaches are commonly found in and around people and food products whether cockroaches are present or not, and there are few studies demonstrating that cockroaches actually transmit these diseases to humans or animals. While it might be prudent to treat serious infestations, it is unlikely that a light cockroach population would pose a serious health risk. Proper food storage and handling techniques that reduce or eliminate pathogens are considerably more important to public health than cockroach control.

The most direct health problem caused by cockroaches is allergic reactions, and again it is heavier infestations that cause the most problems. Many people are allergic to the dust originating from insect or mite cuticle, secretions, and feces. The most serious allergy in homes comes from "house dust," which is primarily derived from tiny mites that are ubiquitous in dwellings. Cockroaches, however, are the second largest cause of this type of allergy, which is most severe in well-constructed homes that minimize exchange of air with the outside world. Common symptoms of cockroach allergies are rashes and respiratory problems, but some serious and even fatal cases of asthma attacks and anaphylactic shock have been recorded.

The potential for disease transmission by cockroaches and the demonstrated allergic responses induced in some people provide a reasonable rationale for cockroach control in many situations. However, the issues with cockroach control are the same as those we encounter every time we attempt to manage pests: When is the pest infestation serious enough to justify control measures? Can we tolerate a low level of this pest? Are chemical pesticides necessary? Are slower-acting nonchemical methods feasible?

When it comes to cockroaches, the thinking of consumers is skewed toward rapid eradication using strong chemicals, even though other, slower acting, environmentally more friendly techniques are available and effective. Tolerating a few cockroaches is not considered an option for most homeowners, and certainly not for public places like restaurants and food fairs. Chemical control is still the major option selected when cockroaches or other insects scurry across the floor.

There is a large arsenal of chemicals that can be purchased by those who want to do it themselves, available at almost any hardware, garden supply, or drug store, and even in supermarkets. These are sold under militaristic-sounding labels like Ambush or Sidekick, or more ominously under their chemical name alone: Diazinon, Malathion, Permethrin, or Methoxychlor. Professional pest controllers have additional, more toxic substances at their disposal, but both the publicly available and the professional insecticides interfere with nerve transmission in insects, and also in people at higher doses.

It is difficult to assess whether widespread use of these chemicals domestically and by the structural pest control industry presents serious health or environmental hazards. Certainly today's pesticides are heavily tested prior to approval, and their use is carefully regulated. Nevertheless, some of the pesticides readily available today are descendants of chemicals originally developed for use in warfare. Although the pesticides that are used in contemporary pest control are considerably less toxic to humans than the earlier compounds, they still can pose a serious toxicological threat if ingested or breathed at higher than recommended doses. Symptoms of pesticide poisoning include abnormal vision, nausea or vomiting, excessive salivation, muscle weakness, respiratory distress, lethargy, and heartbeat irregularities. Even today's relatively benign chemical pesticides can easily be fatal if misused. Clearly these products should be handled with considerable caution.

A cursory reading of the labels that must legally accompany pesticide products provides cause for reflection as to whether they really should be applied at all. Take Siege, an insecticide that is supposed to be used as a spot or crack and crevice treatment for cockroach control and is potent enough to be classified as "only for sale to, use and storage by pest control operators." Its legal label includes such precautionary statements as "may be harmful if swallowed. Avoid skin contact. Wear rubber gloves during application. Wash thoroughly after handling. Do not contaminate water, food or feed by storage or disposal. Use Siege only in areas not easily accessible to children and pets. Do not use in the food/feed product areas of food/feed processing plants, restaurants or other areas where food/feed is prepared, processed, served, stored or exposed." The product also comes with a disclaimer: "The label instructions for the use of this product reflect the opinion of experts based on field use and tests. The directions are believed to be reliable and should be followed carefully. However, it is impossible to eliminate all risks inherently associated with use of this product. All such risks shall be assumed by the user."

This label is typical of chemical pesticides, but of course the consumer and even many professional pest controllers are not sure how seriously to take it. Are these label warnings just legalese to protect the company in the event of the occasional misuse of their product or the filing of an unwarranted law suit, or are there serious intrinsic dangers to using chemical pesticides around the house? This question is much easier to answer for agricultural uses of pesticide, where innumerable studies have shown the health and environmental costs of pesticide application relative to their benefits in pest control. Curiously, domestic pesticide use has not received the same scrutiny, but there are hints in the medical literature that home pesticide use may be causing more problems than one might assume, judged by the ease with which people can buy and use these products.

One way to assess health problems caused by pesticides used at home is to evaluate the number of pesticide-related incidents reported to the network of U.S. Poison Control Centers. This network of 68 reporting centers covers an area that includes roughly half of the U.S. population, and it received close to 2 million calls in 1992 requesting assistance with some type of poisoning incident. Most of these incidents involved accidental ingestion of products stored around the home, and pesticides ranked seventh in importance, after such substances as cleaning materials, analgesics, cosmetics, cough and cold preparations, plants, and bites or stings. About 60 percent of the 70,687 reported home pesticide incidents involved children younger than 6 years old, and included 20 fatalities and about 1,300 incidents that were ranked as moderate to major in medical importance, requiring some type of emergency care.

Other reporting networks indicate similar poisoning problems following standard applications of insecticides commonly used for household pest control. For example, the U.S. Environmental Protection Agency Pesticide Incident Monitoring System reported that 15 percent of all reported poisoning incidents that involved Diazinon, Malathion, Chlorpyrifos, and Propoxur from 1968 to 1980 came from residential settings.

Another way to evaluate the health implications of home pesticide use is to examine statistical data on the incidence of pesticide-induced cancers in residents of homes using these products compared with those that do not use pesticides. Unfortunately, this type of study is difficult to conduct, and the data available are generally unreliable because they were taken from studies conducted for other purposes, in which pesticide use was a side issue. A few studies claim an association between household insecticide use and disorders that are known to be inducible by pesticides, such as aplastic anaemia, acute leukemia, soft tissue sarcomas, and some brain tumors. However, these studies are based on very few cases, on interviewing techniques that rely on people's recollections many years after diagnosis, and on somewhat questionable statistical techniques.

For example, a 1995 study from the University of North Carolina, whose results were widely broadcast by television and print media, suggested that yard treatments with pesticides during pregnancy contributed to the incidence of soft-tissue sarcomas in the children exposed in utero. However, the media neglected to mention that the same statistical analysis appeared to suggest that pesticides *protect* against brain tumors and lymphomas. If we reject the latter conclusion, then logically we should also at least question the legitimacy of the former.

Another approach to evaluating the relation between home pesticide use and health is to test whether pesticide amounts considered to be above threshold levels for human health effects are found after applications that follow label directions. Generally, most broadcast applications of insecticides in homes require residents to be absent for 1–4 hours following application, and suggest ventilating the home prior to re-entry. However, a number of studies suggest that airborne pesticide levels peak at 4–24 hours post-treatment and that both airborne and surface levels can be well above threshold values at that time. For example, one study by Richard Fenske and colleagues from Rutgers University, reported in a 1990 issue of the *American Journal of Public Health*, found that chlorpyrifos residues (trade name Dursban) following an application for flea control were 1.2 to 5.2 times higher than the human No Observable Effect Level (NOEL) for 24 hours post-treatment. As the authors put it, "The dose values derived in this study raise a public health concern. Applications . . . of acutely toxic insecticides may result in dermal and respiratory exposures sufficient to cause measurable toxicological responses in infants." Another study has reported immunological abnormalities in individuals up to five years after exposure to chlorpyrifos, and in another, residents exhibited symptoms of pesticide poisoning five months after Diazinon was applied in their home by a commercial pest controller.

A major problem with using chemical pesticides in the home is the development of resistance by the target insect. Cockroaches can rapidly develop resistance to almost any frequently used chemical pesticide; sometimes resistance to one compound also imparts resistance to other related compounds. The German cockroach is particularly quick to develop pesticide resistance because of its shorter life cycle and greater number of offspring than other cockroach species. The development of resistance means that infested buildings must be sprayed more frequently and with a more highly concentrated solution, which leads to further resistance, or else the pesticide used has to be changed. While rotating a diverse arsenal of pesticides against cockroaches would be prudent chemical management, nonpesticidal control techniques would reduce the problem of chemical resistance to a minimum.

These reports of health problems induced by accidental exposure or deliberate pesticide use, scanty as they are, as well as the development of resistance to pesticides in insect populations, do provide some unease about domestic pesticide use. The level of appropriate concern is difficult to determine but clearly lies somewhere in the middle ground between banning all domestic pesticide use and using these compounds indiscriminately. Perhaps it would be fair to say that the harder chemical insecticides should be used only in more extreme situations, where pest-induced health risks are large and alternative techniques not effective in controlling pests.

Cockroaches are ideal pests on which to practice techniques of Integrated Pest Management that would reduce the use of chemical pesticides in and around the home and in urban environments, since most infestations do not pose serious health problems and could be controlled by alternative methods. The first component of a successful IPM cockroach program should be to know your pest, and the cockroach's characteristics of liking cracks, needing water, and feeding omnivorously provide excellent opportunities for control without normally having to call in the heavy chemical artillery.

A successful IPM program for roaches would begin with pest identification and monitoring of population levels. Each species of cockroach is slightly different in its food and habitat preferences, and proper species identification allows more precise placement of the traps and baits that are important for an IPM cockroach control program. Monitoring can be accomplished using traps with a sticky glue surface placed in and near favorable habitats to check for the presence and level of cockroach populations. This provides better information about when to treat, what techniques to use, and whether treatments are succeeding.

The most important component of a nonpesticidal cockroach control program is simple cleanliness. A cockroach's access to food and grease can be eliminated by keeping food sealed, covering garbage and removing it daily, placing pet dishes in bowls of soapy water to prevent access by roaches, eliminating grease from behind and around stoves, and washing dishes after use. Also, all areas of water accumulation have to be dried up; refrigerator drain pans and house plant soil should be eliminated, pipes with excessive condensation should be wrapped, and leaky taps should be fixed. Premises can be further roach-proofed by sealing cracks and spaces greater than 1/4 inch with caulking or steel wool. Doors and windows should be weatherstripped, and vents covered with a fine screen.

When chemical treatments are needed, baits are preferable because they do not disperse pesticide through the home. Baits such as Max-Force, Combat, and Impact use a variety of food substances housed in tamper-proof containers to attract cockroaches. The food is impregnated with Hydramethylon, a roach poison of low mammalian toxicity that kills roaches 48–96 hours after impact. If a broader treatment is necessary, various dusts can be used in cracks, crevices, joints, corners, baseboards, and spaces where cockroaches hide. For example, diatomaceous earth, a fine powder consisting of dried microorganisms from the ocean, will kill roaches by scratching their waxy outer covering so that they dehydrate and die within a few weeks. Boric acid, a chemical of low human toxicity that is poisonous to cockroaches, also can be dusted into roaches' habitats and may retain its activity for many years if kept dry.

This IPM approach works, and it would seem to be common sense to use it rather than a chemical pesticide attack that may wind up harming the people it is supposed to protect and inducing resistance in the insects it is supposed to destroy. There certainly is good precedent in recent pest management history for the structural pest control industry to adopt IPM. Virtually every other segment of the pest management community has embraced this strategy with almost religious fervor since Rachel Carson sounded the warning bell against indiscriminate pesticide use. Yet we seem to remove our environmentalist hats when we enter the front door of our homes; consumers simply have not been demanding a reduction in home pesticide use.

The offices of pest-control companies are invariably found near but not in residential parts of town, on the nondescript urban side streets that separate homes from heavier industries. They typically are located on the second stories of small buildings; the floors are often covered in worn-out shag carpet, and the rooms are filled with old, inexpensive wooden desks where secretaries take the almost constant phone calls of potential customers. The office walls are frequently lined with dented metal shelving loaded with pesticide products and traps for sale to the few customers who come in person.

On my visits to this dreary domain, I sensed that pest control operators were locked in the past, held against their will in some previous era from which they would like to escape. When I spoke with them about why pesticides are still used so heavily in their industry, I was surprised at how similar their replies were. Each of them stressed to me that they were involved in a service industry. To survive as structural pest controllers, they have to follow the customer's wishes, and customers have extremely low tolerance for pests, demand fast action, and want to see professional exterminators doing something obvious but inexpensive.

One pest manager told me, "When the exterminator says to clean up the garbage and plug holes in walls, restaurant managers say 'Yeah, yeah, just spray or I'll get another company.' " Another said, "It's not what we want to apply, it's what the customer wants." A third told me, "The baits are effective but they're slow. It's hard for an apartment manager to tell his tenant, 'Sorry, you have cockroaches and we'll bait for them, don't worry, they'll be in your cupboards and dishes but they'll be gone in three months.' It just doesn't work that way." Others told me, "The industry has taken a lot of stick for being spray jockeys, but cost is a huge factor here. It's cheaper to send a spray jockey in to hose down the restaurant once a month than it is to put the time into training employees . . . The urban side of things is in the dark ages compared to forestry and agriculture."

The lack of Integrated Pest Management presents a particularly curious paradox in the Pacific Northwest, considered to be otherwise in the vanguard of the environmental movement. British Columbia, for example, is referred to as "lotus land" by the rest of Canada, and is home to every imaginable environmental movement from Save the Whales to Save the Spotted Owl, but not Save Our Homes From Pesticides. This province has seen massive public outcries over the use of pesticides in forestry, agriculture, and parks and has been in the forefront of adopting IPM techniques outside of the structural pest industry. Yet even today it is the pest control industry itself that is leading the movement to introduce IPM techniques into the home, while consumers still demand the rapid gratification of immediately dead pests that only chemicals can provide.

The motivating factors behind the industry's drive to introduce IPM are complex but appear to be rooted primarily in their desire to be perceived as professionals, their genuine worry about the environmental and health impact of pesticides on both their customers and themselves, and their realization that an industry-wide IPM standard would in the end provide more stability and ongoing maintenance work to their contracts.

The public's low opinion of exterminators is exacerbated by the poor level of training currently provided to individuals who enter the industry. Verne Gilpin, President of the Urban Pest Managers Association in British Columbia, has a typical background for a structural pest controller. He was trained as a student in marketing, had difficulty finding a job in that field, and ended up working as an exterminator with a friend. He now owns his own very successful company in the Vancouver area, A-1 Pest Control, and is promoting a new generation of pest managers that he hopes will be much better trained than he was. Gilpin described the typical instruction given today's structural pest controllers by saying that "a good portion of the industry is not well-trained, and there are no regulations to make them well-trained. I can take anyone walking in this door today and within three to four days I can get them to pass the exams and the next day get a service license, with absolutely no practical training."

The implications of poor training were well-illustrated by an American incident that took place in a small town in North Georgia in 1985. A termite exterminator sprayed the outside and inside of a wood-frame house with chlordane, instead of putting it into the ground as the manufacturer's label instructed. When the occupants of the house came home, they did not notice right away that something was amiss. But eventually they realized that the clothes in their closet, the two-year-old's toys, their TV, camera, everything they owned had been covered with chlordane, a highly toxic pesticide that has since been banned for domestic use in the United States. The family, who had to abandon their home and its contents, sued the exterminator for damages, including medical damages because one family member developed chemical sensitivities as a result of exposure. During the trial, the pest-control technician at the site was shown to be functionally illiterate and unable to read the manufacturer's instructions. The jury found in favor of the plaintiffs.

Situations like this should become less likely in the future, because structural pest associations and pest controllers embracing IPM have been lobbying government regulatory agencies to adopt stricter requirements for training, licensing, and licensing renewal. In British Columbia, their pleas to the government have fallen on the very receptive ears of Linda Gilkeson of the Ministry of the Environment, who was recently hired to promote IPM throughout the province.

Gilkeson has an unusual background in many ways, but perhaps is typical of today's generation of pest managers in promoting reduced pesticide use without the extremist approach that often characterizes many public environmental advocates. She attributes her interest in IPM to an incident that occurred when she was a child: "My Dad sprayed flies at the window sill when I was a little tiny kid and killed my goldfish, and to this day I have never got over poor old Robin Hood and Little John floating belly up." This early experience led her to an alternative lifestyle, from growing all her own food organically on a small farm, through a Ph.D. in entomology and employment in a private company developing biological control techniques, to her current life as a government employee charged with the mandate of bringing IPM into our homes and cities.

Gilkeson's major task has been to target consumers with a barrage of IPM promotional materials, while upgrading the training and changing the philosophy of professionals on the structural and urban pest management scene. The current training procedures for entering the structural pest control industry in British Columbia are an IPM advocate's worst nightmare. Potential workers in this industry only need to pass a test based on the aptly named *Handbook for Pesticide Applicators and Dispensers*. This 250-page handbook is chock full of information about pesticides, relevant legislation, safety, pest identification, and application technology, all admittedly crucial information for anyone using pesticides. What is missing, however, is any underlying philosophy that might lead a pest manager toward nonpesticidal techniques. For example, the Handbook contains 30 pages devoted to pesticide application equipment and calibration, but only 15 pages devoted to "Protecting the Environment," and virtually nothing on Integrated Pest Management.

Gilkeson and the Urban Pest Managers Association are about to alter that by changing the way structural pest companies are licensed. Their new guide, *Integrated Pest Management Manual for Structural Pests*, will provide background for an examination that will test on Integrated Management Techniques rather than focusing exclusively on pesticides. If passed, the exam will yield an IPM Certificate rather than the current Pesticide Applicators License. The Handbook will stress monitoring followed by physical, cultural, and mechanical control practices, with pesticides considered only as a last-ditch alternative for more serious and intractable infestations.

The new IPM Handbook will be supported by various changes in licensing regulations that also will improve structural pest controllers' training and ensure that they are aware of new advances in pest control. Renewal of licenses will be required at three-year intervals and will be accomplished by attendance at regular seminars and approved courses that will provide updated information on pest control. Finally, establishment of a new structural pest company will require at least three years' experience in the industry, which will prevent those with no practical experience from working on their own.

Governments at all levels also are beginning to require that IPM techniques be used in structural pest control. At the municipal level, most large cities now include IPM-trained health officers on staff, and health laws are more focused on pesticide reduction. Angelo Kouris of the Vancouver Health Department described the city's philosophy and practice by saying that "we cannot stop people from applying approved chemicals. What we emphasize is that we would like to see the Integrated Pest Management approach. We don't want to see chemicals used unless you have to." Dick Heath of the British Columbia Ministry of the Environment told me that permits for pesticide applications at state or provincial levels "want to see that you're managing pests in a holistic manner, and not depending on chemicals. If pesticides are necessary, show us why."

The U.S. General Services Administration runs a program called Termination of Extermination, and GSA building contracts now state that "there shall be no scheduled pesticide applications, and you must monitor each area with sticky traps and analyze results before recommending treatments."

The influence of government on promoting IPM through its powers to license and regulate pesticide use has been crucial in promoting these new techniques, but public education is viewed as equally important. Gilkeson's Ministry of the Environment office has put out a series of brochures that are being distributed to households, garden shops, and municipalities in British Columbia. Titled *Safe and Sensible Pest Control*, they each focus on a different pest, and each is billed on the front cover as "one in a series of environmentally aware solutions to home and garden pest problems." On the back cover, each brochure says, "You can control most pests around your home and garden without harming the environment or poisoning your living spaces. This series of common sense guides was developed by B.C. Environment to encourage safe, practical alternatives to overuse of pesticides at home. Non-toxic pest control really works; try these simple solutions first."

Perhaps the most significant barrier to implementing an IPM approach in structural pest management is that IPM is expensive. Although there are many tangible benefits to IPM, low short-term cost is not one of them. Albert Greene, an entomologist with the Building Services Group of the U.S. General Services Administration, described the economics of structural IPM to me this way: "There are precious few tangible economic benefits to structural IPM . . . It has become obvious that cost is its greatest Achilles' Heel. Clients always pay more for the premium service of IPM compared to no-frills, hose-em-down extermination. Standard commercial pest control rates in the eastern U.S. are generally about \$30–\$40 per hour. You cannot purchase true IPM service for less than \$65–\$85 per hour, and prices are frequently much higher. Quality costs."

Greene points out that "from the consumer's or policy maker's point of view, structural IPM is desirable mainly because of perceived risk avoidance, public health (especially where children are concerned), minimization of toxic torts, political correctness, etc." In spite of the higher cost, the GSA implemented IPM in over 100 large federal buildings that it manages in and around the District of Columbia. Results have been impressive. Requests for pest control service declined by almost 25 percent, service calls resulting in pesticide applications were reduced from 95 percent of calls to an annual average of 40 percent, and broadcast spraying of hard chemical pesticides like Dursban was virtually eliminated in favor of nonliquid bait products that do not vaporize and have no effect on air quality. The IPM structural pest program is now being expanded throughout government buildings in the United States.

Whether it's cockroaches or fleas, silverfish or ants, the basic tenets of IPM are slowly making inroads into today's home and business environments. The structural pest control industry has been the slowest sector of pesticide users to adopt IPM, mostly because customers have been willing to let—indeed have demanded that—pest controllers in their homes do things they would never condone outside. In addition, IPM is more expensive, and homeowners and businesses prefer cheaper chemicals over safer alternatives. Structural pest controllers have been struggling to become more environmentally friendly, but it is largely the consumer that has limited the inroads that IPM techniques have been able to make in structural pest management.

Historically, a considerable amount of domestic pest control simply has not been necessary, or if necessary was done using more chemical control than might be desirable. Today, we are beginning to move beyond knee-jerk reactions to structural pests and are starting to control them in a more environmentally rational manner. However, the IPM movement in the structural pest industry is still in its infancy and needs to be nurtured. As Linda Gilkeson told me, "Most homeowners have no idea what they're doing for pest control. Most of what they are doing is either done too late or it's done for a problem that's perceived and not real. So much of the structural pest control issues are strictly sanitation, repairing and blocking off entrances, cleaning grease off the surfaces so the cockroaches won't eat it, cleaning out drains . . . When we talk to homeowners and home gardeners, we very much try to do education about nature. Stewardship in pests is the issue, and prevention is the most important tool there is."

CHAPTER FOUR

Weeds

"You can't just let nature run wild."

Walter Hickel, Governor of Alaska (1992)

Every organism has a habitat, a niche with the food, water, and shelter needed to survive. Humans are unusual in being able to thrive in many habitats, and we are unique in the extent to which we can invent our own surroundings. Indeed, we have created an entirely novel ecosystem, the city, which has altered the physical and biological environment to a degree not previously accomplished on earth, at least not by living organisms.

Remarkably, considering what an altered environment cities are, nature frequently thrives in the city ecosystem. But the city version of nature is skewed toward a particular type of organism, one that we call a weed. We usually think of weeds as plants, but if we extend the concept to mean any potentially problematic organism that is found out of its natural place, then nature in the city is full of weeds, from dandelions to coyotes, plantain to rats, clover to geese. A deer, coyote, or dandelion in a country meadow is not a weed, but in the city these and many other organisms become problems, and we have had to maintain an extensive infrastructure of urban pest control to deal with them.

Weeds are not the only form of nature found in cities. We, unlike other animals, tend to foster the presence of other organisms in our habitat, not for food but for the sole purpose of affirming some sense of our place in the natural world. Thus, a well-planned city usually includes forested parks, rights-of-way, expansive lawns, wildlife refuges, zoos, and botanical gardens interspersed between the buildings, sidewalks, roads, and parking lots. This urban version of nature is a hybrid, however, groomed and sanitized for consistency with the urban setting. It is a simulation of the real thing, designed to provide some elemental feeling called "natural" within a highly artificial environment.

Though we attempt to replicate nature in cities, we don't want it to get too close. The squirrel in the park that scurries to pick up a peanut is cute, but let that squirrel nest in our attic and it becomes vermin. At a Monday-morning coffee break we might relate the fleeting view of a deer, raccoon, or coyote we saw while jogging through the park, but when the deer nibbles in our garden, the raccoon bites our kids, or the coyote eats our cat, we're quick to call the city government and demand that it be removed. Our lawns remind us of open spaces, and perhaps we achieve a certain degree of tranquility when we sit on the back porch in the late afternoon, sipping our tea and gazing over our backyard territories. But let a few dandelions or some chickweed invade this manicured version of a meadow and we grab a can of Roundup or Killex and spray it to death.

Our challenge in the urban setting is to create an environment in which we can coexist with other organisms, both because their presence soothes us and because we have little choice, since weeds will enter our cities no matter how urbanized the environment may be. It is remarkable that other species can thrive in our cities, considering how disrupted urban ecosystems can be compared with rural settings. Think about how a city appears to wildlife. Plant communities are diminished, leaving monocultured lawns with reduced cover; food chains are severely disrupted; much of the ground is waterproofed with asphalt and concrete; and urban streams, lakes, and ponds are polluted by runoff from human activities. Even the climate has been modified—city-dwelling organisms are exposed to less sunshine, more clouds, lower wind speeds, and higher temperatures than their rural cousins. The impact of this highly altered ecosystem is to eliminate many species altogether, reducing competition and allowing populations of a few hardy, well-adapted weed species to flourish.

And flourish they have. In Chicago and Winnipeg, white-tailed deer are found at densities of 83 per square mile, up to three times higher than their densities outside of these cities. In Los Angeles, about 400 coyotes have been officially counted within the city limits, but this figure may be much too low: one three-month trapping program in the Glendale section of L.A. caught 55 coyotes in less than a square mile. In Chicago, the 1980 rat population was estimated at 6 million, roughly twice the human population. In Washington, D.C.'s Lafayette Park, adjacent to the White House, gray squirrel densities of 118 animals per acre have been recorded, 10 to 20 times the populations found in rural settings. In Vancouver, hundreds of beavers are trapped out of culverts each year, and municipal authorities receive over 3,000 annual calls complaining about raccoons.

But when it comes to nuisance complaints, birds win the competition wings-down. They are the most common type of wildlife in urban areas, and few people would deny that the trill of a songbird calling at dawn from a bird feeder is one of nature's most eloquent voices in the city. But pigeons, starlings, and geese are another story. Pigeon droppings deface statues and buildings, clog drain pipes and air intakes, and contaminate the air we breathe, the water we drink, and the food we eat. Pigeons carry and transmit numerous diseases, such as encephalitis, histoplasmosis, cryptococcosis, toxoplasmosis, pseudotuberculosis, and salmonella, and their parasites include many organisms that also infest humans, especially fleas, ticks, and mites.

Modern urban pigeons are direct descendants of European rock doves that are thought to have been brought to North America in 1606 as domesticated birds. They have been inordinately successful in city environments because of the abundant shelter and nesting sites provided by buildings, and they thrive by feeding on trash and on the seeds and nuts thrown by misguided urban bird lovers. These highly adaptable birds are difficult to control because they quickly acclimate to most control measures. Thus, traditional techniques such as loud noises, alarm calls, flashing lights, and pyrotechnics are ineffective, besides being unacceptable for use where people live and work.

The Canada goose is another nuisance species whose population has ballooned to pest proportions in our cities. These splendid birds, honking while migrating high in the sky in V-shaped formations, evoke the most primal memories of our human connection with nature. Yet their gray, slimy excretory pellets foul our water supplies and make golf courses and playing fields unusable for people. Geese also are highly aggressive, territorial animals and will chase children and dogs with little provocation. In many areas Canada geese have lost their migratory habits and become resident. In Minnesota, for example, the urban goose population is estimated at 30,000–40,000 resident birds, more geese than the additional 25,000 migrating animals that use urban habitats while in transit.

City parks provide excellent habitats for Canada geese. Fountains and artificial waterflows furnish water, highly fertilized lawn areas supply abundant food, and the openness of parkland provides an essentially predator-free environment for the birds, since they can see attackers coming over long distances. In addition, our major interaction with geese is our warped compulsion to feed them bread and nuts, which exacerbates the conversion of geese from a glorious natural species to a serious pest problem. These "charitable" food donations not only allow bird populations to increase even further, but bread and nuts are nutritionally deficient for birds, and predispose resident birds to contagious diseases that can spread to the more natural migratory flocks.

The Canada goose problem in Vancouver recently exploded to the point that the City Council passed a law levying fines of up to $2,000 on anyone who persists in feeding wildlife. They also spent $10,000 to relocate geese to an environment outside the city and away from the trendy and touristed part of town, False Creek, where bird-feeding had produced a goose infestation. Culling the flock was not an option, since killing birds would have generated an enormous public outcry from bird lovers.

Canada geese are one of the reasons for the recent creation of the

Urban Wildlife Committee, an unofficial consortium of park employees, researchers, wildlife preservation groups, trappers, pest controllers, and humane society volunteers who are concerned about the increasing success of "generalist" species in the city. The chair and current spokesperson for this group is Mike Mackintosh, who manages the zoo in Vancouver's Stanley Park. Stanley Park is internationally recognized by urban park connoisseurs as a gem of city parks. Located in downtown Vancouver, it is one of the world's largest urban parks, with spectacular views of the ocean and nearby mountains juxtaposed with skyscrapers and the Vancouver port. Mackintosh works in a small, woodsy A-frame building nestled within the tall forest at the center of the park. His office is decorated with deer and wolf photos, bookcases filled with works like the *Animal Life Encyclopaedia,* and coffee cups imprinted with wildlife emblems. Thin, fit-looking, and casual in appearance, he is a highly articulate and focused advocate of diversifying wildlife in the city, both to reduce pest problems and as a worthwhile end in itself.

Mackintosh's view of Stanley Park is different from that of most park users. Where they see nature, he sees "an ecosystem designed almost with Canada geese in mind. We're making an open invitation to them to drop in and stick around. I don't think in many cases urban landscape planners have done their homework on it . . . I think we could look a little more seriously at creating habitats that aren't going to be quite as supportive of Canada geese in the future."

The annoying nuisance problems caused by pigeons and Canada geese are only a small component of a much larger issue that concerns how our styles of urban development mold wildlife diversity in the city and create pests out of previously innocuous organisms. As Mike Mackintosh put it, "You take an area like Lost Lagoon in Stanley Park, which appears to be just teeming with life, well, it's teeming with only a very few forms of obvious life at the moment. You can go down there and see hundreds of squirrels and dozens of raccoons and skunks . . . Certainly nearby homeowners that have trouble with raccoons invading their roof or skunks under their porch or squirrels in their kitchen are very unhappy with the whole business. And then there are people who would like to see more wildlife and don't really appreciate the problems that occur with generalist species."

Bird problems in and around cities can escalate well beyond nuisance levels. In the United States each year, 98 million birds die by flying into windows—merely a nuisance to humans—but some fly into airplane engines, occasionally taking the planes down with them. The first recorded fatality caused by a bird colliding with an aircraft was early in the days of human flight. In 1911, Galbraith Rodgers was the first person to fly across the entire United States, making 70 scheduled landings and surviving 19 crashes during the trip. His luck didn't hold, however. He met his death in California the following year in a crash caused by a bird collision.

In 1960 an Electra taking off from Logan Airport in Boston hit a flock of starlings, crashed, and killed all 62 people on board. A 1975 DC-10 flight at Kennedy Airport in New York crashed on takeoff following a collision with several Canada geese; the plane exploded and was destroyed, but fortunately there were no fatalities other than the geese. Kennedy Airport is a particularly hot spot for bird collisions, with over 100 serious strikes per year, an average of five for every 10,000 plane movements. Worldwide, there are about two collisions per year for civilian and corporate aircraft that result in aircraft loss, with an average of six human deaths annually, and the figures for military aircraft are thought to be considerably higher.

Even the highest of high-technology human endeavors can be grounded by birds. In June 1995 a flight of the $2 billion space shuttle *Discovery* was prevented by a few male yellow-shafted woodpeckers that drilled four-inch-wide holes in the shuttle's external coating. These birds normally hammer on dead limbs as part of their courtship and territorial displays, and were trying to attract the attention of females by using large holes banged in the shuttle as an especially monumental selling point.

Bird and aircraft encounters are most acute in coastal cities because airports often share space with major flyways for migratory birds. Coastal airports tend to be located near marshes and wetland areas that provide excellent habitat for both migratory and nonmigratory bird species to feed and nest. And if that wasn't enough of a problem, wetlands and many bird species that inhabit them are strongly protected by both American and Canadian law, so that airport bird managers have to be enormously creative to keep birds out of the flight path of incoming and outgoing aircraft.

One of the more effective airport bird managers around is Roch Grondin, who practices his unusual craft at the Vancouver International Airport. Grondin is a stocky French-Canadian from Quebec who retired from the Canadian military in the mid-1980s and was hired to run the bird control program at the Vancouver airport, although he had little experience in that field. His naivete worked to his advantage, however, because he came at the job with few of the preconceptions that had hindered biologists from controlling birds at the airport. Grondin treated the problem much like a military campaign, but he had to use nonlethal tactics because of environmental regulations and public attitudes. He devised some novel methods that, although simple in concept, have made the Vancouver airport a world leader in bird control at environmentally sensitive sites.

An airport tour with Grondin highlights the contrast between our high-technology modern world and the nature that is increasingly being shoved aside in our cities. He works out of a small, isolated terminal at the edge of the Vancouver International Airport, where float planes take off for the wild interior of British Columbia. Well-dressed business travelers don't make their way here, but sportsmen and loggers wait for their flights in the run-down coffee shop, where Vancouver's caffe latte culture has not yet made inroads. The amenities shop doesn't stock perfume, but you can buy any kind of fishing fly your heart desires, or a full water survival suit if you're not all that confident about flying with a company named Wilderness or Coastal Mountain Airlines.

The Grondin airport tour drives through the grasslands and meadows that surround the runways and stops at the beginning of the main runway, beyond which is some of Vancouver's finest marshland and bird habitat at the edge of the ocean. The marsh is a peaceful and tranquil place in the few seconds of quiet between jumbo jet arrivals, with red-winged blackbirds flitting through the marsh grasses and the occasional heron taking to the air. The airport, located on one of North America's most significant bird migration flyways, hosts over 200 species annually. The marsh itself is strongly protected by Canada's various environmental and wildlife agencies, especially because of its significance for migrating snow geese.

Standing with Grondin at the very end of the North American continent, between marsh and runway, I could see clearly how our attempts to manage nature in the city have produced a hybrid version of the natural world that is almost guaranteed to create pests. To the west, toward the marsh and ocean, is a tiny remnant piece of nature only a few hundred yards wide, a small token of the vast marshes that used to ring the Vancouver area and serve as nesting and stopping grounds for a great cornucopia of bird life. The habitat here is dominated by cattails, sedges, and bulrushes that gradually yield to the extensive mud and sand flats of Sturgeon Banks. The marsh and nearby ocean teem with sculpin, herring, flounder, salmon, cod, marine worms, insects of all varieties, clams, crabs, and voles, food for the avian residents and visitors.

Even in this small, contained remnant of nature, we still see the hallmark signature of the unspoiled natural world: diversity. Within this marsh and adjoining intertidal and ocean areas can be found 310 bird species that nest, rest, or feed at some time during the year, with up to 12,000 individual birds per square mile, the highest diversity and abundance of birds north of California. Here we can see loons, grebes, cormorants, herons, swans, geese, ducks, cranes, rails, coots, sandpipers, plovers, swifts, hummingbirds, woodpeckers, larks, wrens, warblers, waxwings, shrikes, vireos, and many more bird types. Some are international travelers, such as the snow geese that winter here but migrate north to Siberia to breed. Others use the marsh year-round, like the magnificent great blue herons that nest in the marsh during the summers but live and feed here all year long. No single species predominates to the exclusion of all others. Balance is the vision the marsh provides, with a natural ebb and flow of competition and predation, food chains and territories.

To the east is quite a different scene. The airport habitat has been so severely disrupted by runways, terminals, hangars, and vast expanses of closely cropped grass that only a few bird species survive. Here diversity is low, but populations of the few bird species that remain are high and have expanded to the point that they have become pests. Starlings, gulls, crows, and swallows make up the vast majority of the airport's birds, and these are all species that thrive in highly disturbed habitats. If Grondin and his crew ceased their work for even a coffee break, the bird populations around runways would increase to a dangerous level. If the crew went on strike for a week, the airport's pristine safety record would take a very rapid nose-dive.

Bird control here involves an integrated program of habitat management on the airport site itself and frequent patrols to frighten birds away from runways. To begin with, Grondin instituted a simple but effective program of regular grass mowing and ditch drainage to reduce the value of airport land to birds, and also insisted on removing all trees to eliminate bird roosts, since, being a military man, he considers "starlings and other flocking birds to be deadly force for aircraft." Larger birds such as gulls have been mostly eliminated from the airport area simply by tight garbage control: removing most of the food sources that attract gulls to peopled sites.

Grondin and his four-member team also drive past all runways at 5–10 minute intervals and use a variety of loud sounds to frighten birds away. Since different species are frightened by different sounds, they use a combination of sirens and distress calls played from truck loudspeakers, and big bangers and screamers shot out of pistols and shotguns. His team also has convinced airport authorities to put projecting sharp spikes on any ledges and to use an Australian bird repellent paste, Hotfoot, that can be smeared on lights and overhangs.

In addition, Grondin works in close cooperation with the government's Fish and Wildlife Branch. He shares their love of birds, tempered by the pragmatic need to protect birds and humans from each other. Together, Grondin and Fish and Wildlife move nests and birds in the marsh out of the direct flight path used by incoming aircraft. Finally, Grondin has developed a raptor enhancement program, which includes regularly releasing barn owls into hangars so that these predators can feast on birds nesting in the hangar. Ironically, although humans are prohibited from killing birds, raptors are not; government agencies encourage raptor enhancement programs, and Grondin even maintains a small bird hospital at the airport to treat and then release injured raptors.

Grondin's multifaceted program of bird control has been as successful as any in the world, reducing bird strikes at the Vancouver airport to fewer than 20 per year. The few strikes that still occur are less damaging, because the bigger birds that cause the most damage have been largely eliminated from aircraft areas. As Grondin points out, "The $500,000 annual cost for this program is a bargain, considering that even a damaged engine can cost over $5 million to repair."

People may not always appreciate the problems created by bird feeding or poor park design, but at least one pest in the urban ecosystem receives a full measure of attention because of the problems it causes, and that is the rat. There are a number of closely related rat species, with the most significant pests being the Norway and roof rats. But all rat species in the West today originated in Asia and spread to Europe in boats and caravans as trade expanded between the two continents during the Middle Ages. The first rats arrived in North America on boats with the earliest European explorers; roof rat skeletons have been found in excavations from Haiti where sailors stayed following the 1492 wreck of the *Santa Maria*.

Rats are the perfect generalist species and were ideally preadapted to thrive in increasingly dense human settlements. Bobby Corrigan, a rat expert from Purdue University, has called rats "lean, mean gnawing machines, and the number two most successful mammal on the planet, after humans." Their strength is prodigious. Rat teeth are rated at a hardness index of 5.5, greater than iron, copper, or aluminum, and a biting rat can exert a pressure of 24,000 pounds per square inch. Rats also are quick and tough; they can jump two feet high off the ground and are known to survive falls of over 180 feet.

In nature, rats exhibit an enormous flexibility in life style. Rats will eat almost everything but prefer succulent seeds, fresh vegetables, fruits, and grains. They also eat meat, preferably freshly killed or even alive, and will dine on ground-nesting birds, ducks, and even young pigs. If hungry enough, rats will eat other rats. They nest in virtually any hole or crevice but prefer old birds' nests or abandoned animal dens that they can insulate with plant material. Rats also exhibit another common characteristic of pest species, a high reproductive rate. A rat becomes sexually mature at two to four months of age and can breed year-round; a female's litter can reach up to 22 young.

Rats likely began associating with humans as our earliest ancestors arrived in Asia. It was not until the advent of large cities, however, that rats and humans became so closely associated. Asphalt jungles may not be a suitable habitat for many species, but urban environments are prime real estate for rats, and their populations have responded accordingly. Today, from the point of view of humans, this cosmopolitan species has become the most serious of urban pests, causing incredible damage to our health and buildings.

The noise of rats walking, climbing, and fighting behind walls and under floors is one of the most terrifying of sounds, which is not surprising considering the problems associated with rats. For example, rat bites are frequent in decaying inner cities, where high rat populations emerge at night and attempt to feed on people, especially infants. One report in a 1979 issue of *Newsday* described how rats had feasted on the legs and fingers of a living indigent man for two weeks in the basement of a Chicago tenement, until alarmed residents finally heeded his screams and rescued him. Rats will chew on more than people; a circus owner in Germany was forced to kill three trained elephants because rats had gnawed their feet, and rats were seen eating the hides off living alligators in a Los Angeles alligator farm.

Rats transmit myriad human diseases. Some, like plague and murine typhus, are vectored by fleas that live on rats but also feed on humans, thereby transmitting the disease. Plague has caused some of the most serious epidemics in the history of human disease. Over 11 million cases were reported in Asia from 1898–1923, and in this century there have been plague outbreaks in San Francisco, New Orleans, Galveston, and Los Angeles. In 1995 an outbreak in central India killed hundreds and resulted in international quarantines until the problem was brought under control. Other infectious diseases transmitted by rats include jaundice, rat-bite fever, food poisoning caused by bacteria carried in their feces, and hantavirus.

Rats are responsible for considerable physical damage to buildings and infrastructure. Their gnawing habits lead them to chew on and through electrical wires and gas-laden pipes, causing fires and explosions. About half of the reported fires of "unknown origin" are thought to result from rat damage. Rats have been implicated in shutdowns of computer and phone systems by gnawing through data-carrying wires behind walls and under flooring.

Rats are a professional pest controller's dream organism. Unlike cockroaches, even a low-level rat infestation is dangerous. Rat control is mandated by law in virtually every municipality, so that landlords, restaurant owners, or office building managers must hire an exterminator on a regular contract to prevent rat problems, or deal with rats when they appear. Further, rat control requires extensive and routine surveys, diverse techniques, and continual vigilance, so that a rat control contract provides both long-term income and a challenge for pest managers to outwit this successful and intelligent pest.

Rat control is not for the faint-hearted. Animal rights advocates need not apply for these jobs, because killing rats is a gruesome if necessary task. Of course, rodent-proofing buildings, known as "rat-stoppage" in the trade, is the first line of defense. Reducing dead space between walls and floors, blocking all access holes with concrete or sheet metal, storing food in rat-tight containers, and providing tight lids for garbage cans and dumpsters will keep many rats away from people and buildings. The ones that make it through these perimeter barriers can be killed only by bloody techniques.

The main method of rat control used by exterminators, anticoagulant chemicals, is indeed "bloody." The first and still-used anticoagulant was warfarin, also known as Compound 42. This or other odorless and tasteless anticoagulant chemicals are placed in bait stations near rat habitat and are consumed by rats. Then, as the *Handbook of Pest Control* puts it, the chemical "kills by destroying the coagulating powers of the blood and by causing capillary damage. The stricken rodents die a peaceful death from internal bleeding." Chemicals such as warfarin also will kill dogs, cats, and people in a similarly "peaceful" manner, so bait stations must be carefully placed and well-sealed.

Surprisingly, rats can become resistant to anticoagulant chemicals, and an arms race similar to insecticide development has stimulated the chemical industry to produce new anticoagulants as old ones become ineffective. Today's second-generation anticoagulants require only a single dose as opposed to the chronic exposure needed to kill rats with earlier chemicals, but the more potent nature of these single-dose compounds means that pest controllers must be exceptionally vigilant in maintaining bait stations.

Rats can be trapped, but their deaths in traps are no more pleasant. Live traps that attract rats to enter a tube into a cage, but do not permit retreat, are the most commonly used type of trap. These traps are checked regularly, and when rats are found the traps are immersed in water to drown the rats before they are removed. Large snap traps like the common mouse trap also are used, but their nicknames of "guillotine" or "break-back" traps say something about how pleasant the experience is for the rat. Another type of trap is the glue board, developed for mice, which attracts rodents with baits, then immobilizes them by powerful sticky glues spread on the floor of the trap. A lingering death from thirst while stuck to glue probably isn't much fun, either, although rats can escape from these glues more easily than can their smaller mouse cousins.

Urban rat populations are sustained in part by uncovered trash, food thrown on the ground, or bird feed left unconsumed in feeders. Much of this garbage and bird food ends up being consumed by rats. In turn, rats provide an abundant food source, and often the major one, for another urban pest species, the predatory coyote.

The coyote is another of those animal species that trigger both sentimentality and wrath. The lonely howl of a coyote is synonymous with wilderness, especially for people living in western North America, but the numerous coyotes in the city evoke a very different reaction. Listen, for example, to the voice of Sandy McCormick, an irate Vancouver woman and school trustee who wrote a letter to a Vancouver daily newspaper in August 1995 following the death by coyote of the neighborhood cat, George. Titled "Killer coyotes stalk prey on city streets," her letter went on to say, "Coyotes are hungry, desperate serial killers who we know will kill again. They may, however, not be choosy about the number of legs on their next victim. A kindergarten student hunched over in the grass would not seem much bigger than George from a distance. How long before they pick a small child as their prey . . . George had a lot of love left to give when he was taken from us, and I know he has lots of friends in kitty heaven. But the rest of us live in fear, wondering who will be the next to join him."

There is an opposing attitude to coyotes in the city, however, that is eloquently advocated by Kristine Webber, a graduate student at the University of British Columbia who is studying urban coyotes for her Master's thesis research and is a member of the Urban Wildlife Committee. Kristine works out of a decaying castoff trailer with peeling plywood walls and old wooden desks that is parked behind the modern forestry building at the university. Her surroundings reflect the importance ascribed to urban coyote management relative to forestry in British Columbia, but it doesn't seem to bother Webber. She is vivacious, enthusiastic, and very public-oriented, and views herself as a supporter of wildlife in the city. She became interested in urban coyotes from her previous job working in a local veterinary clinic, when customers would express concern about their pets being eaten.

Webber respects coyotes, and believes that we, rather than they, should adapt.

> I think it's really neat to get a glimpse of them. There is intrinsic merit in urban coyotes. This is an animal that's done very well in cities, a setting that many wild animals haven't been able to exist in; kudos to them. We can't expect them to modify behavior that is innate and survival oriented, but what we can do is modify our behavior somewhat. What we need to control is ourselves. We need a bit of a paradigm shift in philosophy and who is responsible for what, and what place urban wildlife has . . . I admit I don't want my cat eaten by a coyote. I mean, I don't. My cat stays inside. I made the choice.

Coyotes do exceptionally well in disturbed environments that provide a mixed mosaic of habitats, such as cities. They fit the typical profile of a generalist species that urban environments support: broad eating habits, flexibility in den sites, high reproductive rate, and intelligent avoidance of humans. Urban coyotes will den in hollow trees, under porches, in abandoned houses, or in brambly thickets, and roam freely through parks, ravines, culverts, and shrubby backyards. They usually mate for life, and a coyote couple can raise up to 15 young in an annual litter. They are rarely sighted by people in the city, except for the most fleeting of views, because they are extraordinarily stealthy in their movements. Even Webber, who spends most of her days and nights looking for coyotes, has only encountered them a half-dozen times in the city.

Rodents are the main food for coyotes, in both city and country habitats, but they also eat compost, insects, fruit, refuse, rabbits, raccoons, and yes, occasionally a pet cat or dog. However, as Webber points out, the most significant urban pet predator is the automobile, which kills many more cats and dogs than coyotes ever will; yet there certainly is no movement to trap and remove cars from the city. There have been attacks by coyotes on young children, but these attacks are exceedingly rare. The last reported fatality due to an urban coyote attack was in 1981, when a young girl was killed by a coyote in Glendale, California.

Animal attacks by another "weed" mammal, the raccoon, are not so rare. Raccoons like to nest in urban attics and basements, creating an awful stench from their urine and feces, and they bite when disturbed. Human activities have dramatically increased urban raccoon problems in two ways. First, the reduction of wilderness habitat near cities has forced raccoons to migrate into towns and cities to find food and shelter. Higher urban raccoon populations have resulted in a greater frequency of attacks on people. The second, and more alarming, problem began when hunters transplanted raccoons from southern U.S. habitats into the mid-Atlantic states, in order to increase game populations for raccoon hunting.

In 1977 a southern variety of raccoon was found in West Virginia that carried a strain of rabies previously unknown in the north, and this virulent rabies strain made its way into New York State by 1990. Since then, the incidence of rabies exposure has risen dramatically. In 1989, 84 New York residents required antirabies inoculations; in 1993 the number was over 3,000, at a cost of over $4.5 million—not trivial. Several rabid raccoons have been found in each of the urban heartlands of New York City, Washington, Baltimore, Boston, Philadelphia, and Newark, and the migration of potentially rabid animals into cities does not bode well for rabies control.

Coyotes and raccoons are by no means the only large animal pests that inhabit our cities. A few other species exhibit the same ability to thrive in altered urban habitats, and they each cause their own brand of mischief. Beavers roam our city culverts by the hundreds, chewing down trees and blocking drainage. Squirrels destroy wooden shakes and shingles and interior insulation as they chew their way into home nesting sites. Even the cutest of animals, such as deer and rabbits, can become pests by eating garden produce and vegetation.

❧

Although all of these large mammal species meet the broad definition of weed, it is plants such as dandelion, chickweed, plantain, lambs quarters, thistles, morning glory, and many others that are most commonly thought of as weeds. Here, also, what is a weed and what is desirable vegetation is largely a product of our personal and societal decisions. However, we have upped the ante for a plant to be desirable in the urban setting by the invention of a highly simplified new habitat, the lawn. Lawns are pervasive in our cities, in backyards, parks, golf courses, rights-of-way, and cemeteries, and we require them to be homogeneous, weed-free environments. Even other grasses are considered weeds in lawns; woe to the emerging stalks of fescue in a kentucky bluegrass planting.

The creation of turfgrass habitats in the city provide an excellent example of how diminished diversity in an urban environment can create pest problems, in this case the invasion of generalist plant species that can thrive in lawn habitats. We desire lawns for many reasons, including visual appeal, fields for sports and golf, and ease of maintenance, but the impact of turf in our cities goes well beyond our human use of these habitats. Lawns require enormous input from fertilizers, herbicides, and lawn mowers to keep them in the weed-free state we demand, and severely diminish the available habitat for other plants and for wildlife.

Up to 60 percent of land in most cities is covered by turfgrass. In the United States, about 50 million acres are blanketed by turf, about 2 percent of the entire U.S. land mass. Over half of this area is found in lawns surrounding homes, an area approximately equivalent to that of all five New England states. This turf is not without its environmental benefits, at least according to statistics compiled by various turfgrass associations. They claim that a single front lawn has the cooling effect of 10 tons of air conditioning, about 2.5 times greater than an average home air conditioning unit's capacity. The same lawn releases enough oxygen annually for a family of four, and the turfgrass in the United States traps 12 million tons of dust and dirt each year. Turfgrass also is valuable as an erosion control plant in roadsides and steep yards. Finally, turf advocates like to point out that the aesthetic value and significance of grass for sport fields and golf courses is impossible to calculate.

Perhaps the greatest problem caused by lawns is the high herbicide input required to keep grass free of weeds. Our societal compulsion for weed-free, perfect lawns, especially around homes and golf courses, is not mandated by health concerns, since weeds in lawns do not transmit human diseases or decrease agricultural food production. Bob Wick, the director of the Western Canada Turfgrass Association, described the impact of this weedless mandate on turf management: "To be competitive, the sod industry has been trapped into a position of having pretty much clean sod, as far as visible weeds, because of the demands of the public. They don't want to buy sod that has weeds . . . The control of weeds is pretty much herbicide-dependent if you're going to set 100% weed-free as the threshold."

Herbicide-dependent turf care for home lawns and golf courses can involve extensive and diverse chemical inputs. Approximately 13 million home lawns in the United States are treated each year with one or more of the 76 herbicides that are available to eliminate weeds. A typical chemically based weed management program begins on new turf by fumigating the topsoil, followed by an application of a broad-spectrum herbicide that kills virtually all foliage; then another herbicide might be applied after planting to kill annual grasses such as crabgrass and foxtail. Finally, broadleaf weed killers are applied annually to maintain picture-perfect, grass-only lawns.

This heavy use of herbicides is largely unnecessary, since healthy grass is tough, resilient, and able to out-compete most weeds if cultural practices are properly done. Professional lawn care experts maintain that home lawns could be largely weed-free by nonchemical management techniques, such as mowing frequently but keeping grass at a relatively high height, leaving clippings on the lawn as fertilizer, watering deeply but not often, aerating mechanically and removing thatch every one to two years, and manually removing the few weeds that remain before they go to seed. Most North American cities have responded to public pressure to reduce pesticide use by avoiding almost all herbicide treatments on playing fields and municipal lawns, and have had few problems maintaining healthy turf, in spite of these surfaces receiving much heavier use than backyard lawns. Evidently, higher standards regarding pesticide use are demanded from our urban governments than what we expect in our own back yards.

Lawns in cities are coming under broader scrutiny as at least some urban dwellers have begun to question the fundamental rationale behind feeding and caring for lawns. Perhaps the biggest question about lawns is not how to manage them, but why have them at all? Some lawn use is certainly justified, but there is an alternative vision of urban environments that would maintain more diverse systems.

Wallace Immen, a science reporter for the *Toronto Globe and Mail*, described the antilawn movement well: "While emerald-green turf remains a status symbol around homes, a grass-roots movement is challenging the multibillion-dollar preoccupation with the care and feeding of lawns. Environmental groups and activists are beginning to persuade homeowners and governments that grass lawns waste resources, and that because they are monocultures, they are more susceptible to pests and infections. Power mowers pollute the air, and fertilizers and pesticides get into the water supply. Natural lawns and gardens are the best way to reduce pesticide use."

Governments also are getting into the act, with programs such as British Columbia's Naturescape, designed to fulfill the dual, compatible objectives of diversifying nature in urban environments and reducing the dominance of pest species. Kits have been prepared to encourage homeowners to create minihabitats in their backyards, with different habitats designed to enhance native vegetation, insects, larger mammals, or whatever the homeowner desires. The reduction of manicured lawns is one objective of the program, replacing them with natural gardens of diverse plantings that don't require pesticides or human maintenance. Susan Campbell, whose job is to implement the Naturescape program, says that it "stresses the primary role of habitat in conservation, and the interrelationship of all living things . . . Naturescape encourages urban residents to create, encourage, and enhance wildlife, whether you have a balcony, patio, backyard, or large acreage. Even corners count."

This concept of allowing a more authentic version of nature in the city is a radical one for urban planners and dwellers, but the movement is growing in support and infrastructure, through groups such as Naturescape and the Urban Wildlife Committee. These organizations believe that pests in cities are a product of our highly disrupted urban habitat, and that a diverse, more natural urban setting would dramatically reduce problems caused by weed species such as rats, coyotes, raccoons, geese, and the many plants that infest our monocultured lawns.

They go further, however, in stressing the intrinsic value of reconnecting urban dwellers to a more realistic perception of nature. The problem with urban environments is not that only a few animal or plant species are dominant. It is common and "natural" for an ecosystem to have a few very populous species. For example, even within recent human history we can recall vast herds of buffalo roaming the Great Plains, dense schools of salmon spawning in our rivers, extensive fir, pine, or spruce forests blanketing the land, and seemingly unending flocks of birds winging south for the winter.

What is unnatural and worrisome about urban environments is that they lack one defining characteristic of healthy environments, diversity. In that way, cities differ from rural ecosystems that might appear to be dominated by a few populous species. City habitats have been molded by human construction and needs, and leave us with a sense of being disconnected from settings in which the dominant influence is not human. In addition, one effect of our supplanting the slower, nature-driven mandates of natural selection has been the dominance of weed species in our cities, and a consequence of urban development has been the transformation of some previously innocuous species into pests.

A new environmental ethic is emerging toward more diversity and less human influence in urban landscape management, and this philosophy is beginning to influence how we design cities. The techniques for more environmentally balanced urban design are simple: if we provide wildlife corridors, unmanaged green spaces, and diverse habitats, urban nature will diversify and pest populations will diminish. Mike Mackintosh described this developing perspective:

> There are plenty of people that talk for humans; animals and park species are my consideration. We have to do something to insure that we retain an appreciation of biodiversity within urban centers. We are primarily urban now, and if you're living in an urban center you're cutting yourself off from the more natural experience, the appreciation of nature and its complexity. It's of paramount importance that we institute programs within urban centers to try to insure that people don't lose their root, so to speak. We [the Urban Wildlife Committee] are trying to turn the tables and show people that living with wildlife around you is more of a blessing than a curse, and people should recognize that. Why would you want to live in a barren concrete suburb? The idea that there should be wildlife corridors and green spaces for wildlife is an idea that I personally think is absolutely essential for the city and its health. We end up in city situations building parks that are in large measure uniform and simplistic, and basically they don't support a wide variety of species. What we have to try to do in urban centers is promote a more diverse population of wildlife species. We have to work against the interests of some of these generalist pest species.

This new generation of urban wildlife managers has a different perspective on what is a pest, and why. They are strong advocates of the rights of animals and plants to have habitats, or at least co-exist in ours. Susan Campbell, for example, promotes Naturescape by pointing out that "stewardship of land requires care, nurturing and protection of plants and wildlife, and respect for the wildness of animals. It's an ongoing and ethical responsibility. Consider your neighbors, and have a sense of respect for wildlife and co-existing with wildlife. There's no such thing as a pest animal. It's a label we give it; it's really a conflict in habitat use."

Diversity, conservation, stewardship—these trendy terms for environmental preservation are not applicable just at some distant wilderness site. City dwellers, too, can serve as stewards of a more diverse and natural urban environment, but part of the growing urban conservation movement must include an increasing tolerance for what we think of as pests. While some pests, such as rats, need to be controlled, others like coyotes, deer, and dandelions are nuisances at their worst, and at their best they provide us with a view of nature within the city. Many people are beginning to believe that they should be cultured rather than condemned, nurtured rather than poisoned.

Perhaps there is at least one major urban species we should consider a pest. As Kristine Webber said,

> I don't think the city should be this sterile place that's only for us, and we should have a big fence around to keep everything else out . . . It goes from one extreme to another. It's fine so long as wildlife is at arms length, but as soon as it impacts anyone personally, it becomes a different issue. A small amount of plants and animals in the city is OK, but when it's a large amount it's unmanageable, and they become pests. We should apply the same ideas to ourselves; a small amount of us, we're fine, when we congregate we're pests. I don't know why we can't view ourselves in the same light that we're viewing all the other species. We fit the pest category a long time ago.

CHAPTER FIVE

The Worm in the Apple

"The codling moth is the 'worm in the apple.' The goal of the Sterile Insect Release (SIR) Program is to prevent this damage by eradicating the codling moth, the B.C. tree fruit industry's number one pest. The SIR Program is really an insect birth control program on a massive scale!"

Brochure, Okanagan-Kootenay SIR Program (1995)

The scenic Okanagan Valley in southern British Columbia is home to one of the most unusual pest management programs being conducted anywhere in the world today. The object of this massive operation is to eradicate a tiny but devastating apple pest, the codling moth, by releasing sterile male moths. This technique of sterile male release has been used before in pest management, although it is rarely implemented because of its expense. What is most unusual about this program is that local urban taxpayers joined with rural orchardists to pick up the steep tab for this high-technology pest control operation, because of their mutual interest in reducing pesticide use in orchards.

What is not unusual about this program is that it is beginning to fall apart. Operational, economic, and political problems have combined to disrupt what seemed like a paradigm for the future of pest management. Instead, the Sterile Insect Release Program for codling moth is becoming a classic demonstration of the difficulty that well-intentioned alternatives to pesticides can have in making it in the commercial agricultural arena.

The Okanagan Valley, which extends southward into the important orchard regions of western Washington State, has long been the major fruit-growing region in western Canada. The Canadian portion of the valley lies about 300 kilometers east of Vancouver, on the other side of high and heavily forested coastal mountains that trap most of the moisture coming off the Pacific Ocean. Little of this moisture reaches as far inland as the Okanagan, so that the valley is hot and dry in the summers and cold in the winters, perfect fruit-growing conditions. The bottom of this thin, 200 kilometer-long valley contains a series of deep lakes that provide irrigation water for the orchards and recreation for vacationing urbanites from all over Canada and the western United States, who turn the Okanagan Valley into a playground each summer.

The region's pleasant weather, beautiful scenery, estate wineries, and quiet lifestyle also have attracted retirees and urban refugees from city living. The Okanagan has proven so attractive that a number of major urban centers have developed in recent years, and the land base for the traditional orchard way of life is rapidly disappearing into urban and suburban developments.

Still, in many isolated areas of the Valley, rows of cherry, apple, pear, and peach trees dominate the landscape. The orchards are arranged on the plateaued tableland above the lakes, and there is no more beautiful or fragrant scene than the colorful spring bloom when the many acres of orchard trees come into flower. In summer and fall, when the trees are laden with fruit, the Okanagan Valley provides urban vacationers with a glimpse of a rural farming life that most of them know only through movies or television.

The orchardists' perspective is somewhat different from that of the tourists, however. Growers' visual cameras zoom in to close-up scenes, where they see their apples being consumed by insect pests, and they smell another odor on top of the floral scents—the acrid smell of chemical pesticides that often are necessary to get in a commercial crop.

The pest that puts growers on their tractors to fog their otherwise idyllic valley with pesticides is the codling moth. This is the key insect in apple pest management, the one that determines how other pests are controlled, and the one that in the end will make the difference between a profitable year or a trip to borrow yet more money from a sympathetic banker. The codling moth is not only an Okanagan Valley pest; it is the most significant pest of domestic and wild apples world-wide and does considerable damage to other crops as well, such as quince, walnut, apricot, plum, peach, and nectarine. Codling moth larvae prefer apples, however, and the proverbial phrase "worm in the apple" was coined for this insect.

The drab gray-brown adult moth is only 1/2 inch long. Its scientific name is *Cydia pomonella,* but its common name of codling moth came from its habit of infesting green English cooking apples called "codlings." Historical records suggest that these insects evolved somewhere in western Asia, eastern Europe, and southwestern Siberia, possibly in and around the Himalayas. Like many other pest insects, the codling moth has spread along with trade and shipping, and it currently infests most temperate fruit-growing regions around the world.

The first adult moths emerge each year when the apple trees are in full bloom, having spent the winter as full-grown larvae in cocoons concealed beneath tree bark. The adults live for 14–21 days, during which time they mate and then lay 30–40 eggs on leaves, twigs, and developing fruit. The hatched larvae burrow into the fruit, leaving a small entry hole called a "sting." They then tunnel through the fruit pulp to feed on the seeds and core. This feeding habit is unusual, because most insects can't stomach the high level of cyanide found in apple seeds. The codling moth is able to digest the seeds because of specialized enzymes in its gut that can detoxify cyanide, thereby overcoming this usually effective chemical defense. The larvae leave the fruit when they have finished feeding, spin cocoons under bark or debris on the ground, and emerge as adult moths a few weeks later. The moths go through one to three generations a year, depending on climatic conditions.

The damage done by a codling moth larva to an apple is not simply cosmetic, although even the presence of a sting will downgrade the apple and reduce the price. A successful larva in an apple will leave the core black and disfigured and produce externally visible exit holes coated with excrement. The moth also spreads easily from tree to tree, although it is not a strong flyer. The practice in commercial orchards of shipping bins of apples long distances to packing houses provides an easy transportation system to move moths from one area to another. In addition, abandoned orchards and unsprayed backyard plantings serve as reservoirs from which the moths can repeatedly recolonize sprayed orchards.

The economic damage caused by the codling moth depends on the number of generations it goes through each year. Regions like Nova Scotia, with short, cool summers and only one annual generation, will experience a 6–10 percent crop loss under insecticide-free conditions. In New York, with 1–2 generations of moths each year, an unsprayed orchard will have 7–35 percent of the apples infested. In warmer regions with longer summers, such as the Crimea, Australia, and the Pacific Northwest, orchardists experience 2–3 annual generations of the pest, and can expect 65–100 percent infestation in unprotected orchards.

The damage level that growers consider serious enough to justify spraying is only 0.5–1 percent infestation. At this level the cost of spraying becomes cheaper than moth damage, especially because the difference between profit and loss can be as low as one out of every 100 apples in this marginal industry. Most significantly, however, a 1 percent infestation rate will rapidly escalate into total crop loss within one or two codling moth generations, so that failure to control the moth at low levels is not an economically viable strategy. For these reasons, growers in the Okanagan Valley and elsewhere in the world will put on three to five pesticide sprays each season.

In the Okanagan Valley, the cost of spraying is over $1 million a year. Even with this high and expensive level of chemical applications, the moth still is responsible for about $2 million in annual crop losses out of a total annual production valued at $40 million. Yet without pesticide sprays, there would be no commercial apple industry in the Okanagan Valley or anywhere else in the world. The economic significance of the codling moth determines how apple pest management is conducted, and control programs for most insect pests of apples are organized according to the need to manage this key pest.

The high frequency and spray volume required for codling moth control has led to pesticide resistance in both the codling moth and secondary pests such as mites, aphids, leafhoppers, leafrollers, and scale insects. Orchardists began this century using one to three sprays of the highly toxic arsenical compounds, which escalated to four to five sprays within a few years, then seven to eight due to resistance. The next generation of compounds was no different; DDT was introduced in the mid-1940s and was effective for a few years, and then resistance developed. The same pattern was seen for parathion and other chemicals in the 1950s–1970s. Today, growers rely on basically one chemical, the organophosphate Guthion, which is one of the most hazardous pesticides licensed for use today. Growers have had to increase the number of Guthion sprays applied over the last few years, similar to previous pesticides, and signs of resistance are beginning to appear in California and elsewhere.

Considering the economic significance of codling moths world-wide, and the heavy pesticide spraying that has been necessary to control it, it is not surprising that the codling moth has served as a testing ground for many novel pest management strategies. The litany of attempted control strategies reads like a history of alternative pest management—banding trees to prevent larval passage to the ground, using overhead water sprinklers to discourage moth flight, releasing numerous predators and parasites, penning hens into orchards to scratch the ground and eat larvae, and attempting propagation of viral and bacterial diseases. Growers have even tried to visually inspect and remove individual infested apples, an incredibly labor-intensive and expensive undertaking. None of these nonchemical techniques have proven successful in commercial orchard management, and chemical spraying has remained the dominant method of codling moth control.

Continued failures of noninsecticidal methods have not discouraged pest managers from trying to develop low or pesticide-free management strategies against the codling moth. The search for alternatives is driven by both public concern about environmental health and growers' distaste for the nausea, vomiting, and other health effects that follow Guthion sprays. In addition, the public and growers are united in preferring "organic" orchard produce, consumers because it fits their healthful image of fruit and growers because organic, or at least pesticide-reduced, can mean a considerably higher selling price.

Today, a mega-experiment is going on in the Okanagan Valley to explore yet another alternative to pesticides: Sterile Insect Release. This technique is theoretically sound, but in practice the outcome of this ambitious program has been decidedly negative.

Sterile Insect Release (SIR) is one of the more interesting pest management techniques that have been developed in this century, but also one of the most difficult and expensive to implement. SIR programs involve sterilizing male insects, usually by irradiation, and subsequently releasing the sterile males to mate with wild females; the result, if the males find females, is a failure to produce offspring. The concept of Sterile Insect Release was tested first on the screwworm, a fly that lays its eggs in open wounds on livestock, especially cattle. The larvae feed on the festering wounds, causing damage to hides and meat and often the death of the host animal. This little fly can cause big damage; livestock losses fluctuated between $70 million and $120 million annually in the southern United States prior to the implementation of SIR.

In 1937 an entomologist named E. F. Knipling developed a method to sterilize male flies by irradiating them with Cobalt-60. Knipling reasoned that, if massive numbers of male flies could be reared, irradiated, and released, the number of fertile females in the wild could be reduced to zero within five generations. His calculations suggested that a ratio of about ten sterile males for each wild female would be required to eradicate the flies in that time span.

After a few failures and some small successes, the first operational sterile male release program was conducted on the island of Curacao, off the Venezuelan coast, where the screwworm was a major pest of goats. In 1954 the U.S. Department of Agriculture and its counterpart Venezuelan agency reared and released 400 sterilized males per square mile per week over a four-month period, which covered four to five screwworm generations. It worked; the screwworm was totally eradicated from the island.

Emboldened by this success, the USDA and southeastern U.S. cattle ranchers implemented a larger program designed to eliminate screwworm from Florida, Georgia, and Alabama. In 1958, 50 million sterile flies were produced each week, and over 2 billion released over an 18-month period. The flies consumed over 40 tons of ground meat each week in the rearing facility, and 20 aircraft flew almost daily releasing the irradiated insects. Again, the program succeeded in eradicating the screwworm from the southeastern United States, at a one-time cost of $10 million but with an estimated benefit of $20 million each year following the program.

The SIR screwworm program then was extended to the southwestern United States, and a facility was constructed in Mission, Texas, that could rear 150 million flies per week. This program also was successful, although it was necessary to keep releasing sterile males for about 20 years. There were some alarming increases in screwworm populations in the 1970s that eventually were attributed to inadvertent production of less competitive fly strains, but the introduction of fresh genetic material into the facility seemed to overcome that problem. Here, too, the economics were favorable, with an annual $5 million budget justified by the estimated $100 million per year savings in cattle loss attributed to reduction of screwworm populations.

Finally, a facility was opened in Mexico in 1976 that could produce 500 million flies per week, and the screwworm has more or less been eradicated in Mexico as well. Economically, this program has been a real winner, with the benefits of screwworm control estimated at ten times the cost of the program.

There have been three other successful Sterile Insect Release programs around the world. In Okinawa, melon fly was eradicated after the release of 50 billion flies, and at a cost of $110 million. In the United States the pink bollworm, a cotton pest, has been prevented from spreading into the San Joaquin Valley, largely via releases of 100 million sterile insects per year, and SIR programs have been a factor in preventing the Mediterranean fruit fly from becoming established at various locations around the world.

In spite of these dramatic and notable successes, Sterile Insect Release is not considered a viable approach for control of most pest insects, for a number of reasons. The primary one is expense. These programs are enormously costly because they need very large, complex rearing facilities, as well as extensive state, country, or even continent-wide distribution and release systems for the sterilized insects. Expensive as the successful programs have been, SIR programs for other insects may be even more costly. For example, operational costs for a program to eradicate the cotton boll weevil by sterile releases would run over $1 billion in the United States alone.

Further, rearing large numbers of any insect can be problematic because diseases can quickly become epidemic in a rearing facility. In addition, many insect species reared on synthetic diets are not competitive with wild insects that have grown and matured on natural diets, and irradiation or other sterilization techniques frequently impair mating competitiveness compared with wild males. Finally, even a short "break" in Sterile Insect Release due to weather, rearing problems, or other technical difficulties will allow the wild population to rebound, and the program must then start anew to reduce the pest population.

Nevertheless, the success of the screwworm program did inspire research into using sterile release techniques against other insect pests, including the codling moth. Agriculture Canada scientists working in the Okanagan Valley during the 1970s became interested in testing SIR against codling moths and conducted an extensive pilot program in the Similkameen region from 1976 to 1978. This south Okanagan area appeared ideal for such a project, because it was isolated, contained a dense concentration of commercial orchards, had relatively few backyard plantings or abandoned orchards, and had excellent cooperation from growers in the region interested in alternatives to pesticide sprays.

Initial calculations indicated that a ratio of 40 sterile male moths for each wild female would be necessary to achieve effective control at the grower-acceptable level of 0.5 percent infestation. To achieve this 40:1 ratio, it was necessary first to reduce the starting moth populations by using chemical sprays and removing neglected trees. Then, sterile males were released over a 500 hectare area in each of three years, spread from May to September. The biological results were generally successful, with damage exceeding the economic threshold of 0.5 percent in only 7 of 436 orchard plots tested. Further, the majority of plots had no codling moths present at all by the end of the experiment.

Although this pilot project demonstrated the biological potential for codling moth control using Sterile Insect Release, the economic parameters were not as promising. The cost of the SIR program was $225 per hectare each year, compared with annual pesticide costs of only $95 per hectare. Further, similar experiments in the state of Washington by

USDA scientists were not as biologically successful. In their project, they could achieve only a 5:1 ratio and were not able to reach a control level that was acceptable to growers. Even the more successful Canadian scientists were appropriately reserved about the practical potential of their results. M. D. "Jinx" Proverbs, head of the Canadian project, wrote that "excellent codling moth control can be achieved in British Columbia apple and pear orchards by SIR, but the method is presently considered too expensive for commercial use . . . Orchards with even 0.5 percent injury at harvest, a level that is accepted by most commercial growers, would require too many sterile insects to make this method of control economically feasible."

Those results normally would have spelled the end of SIR for codling moth control, and indeed the U.S. community did not pursue SIR any further. However, an unusual congruence of political, social, and scientific factors converged in the Okanagan Valley during the 1980s to resurrect the SIR concept and led to the implementation of a full-scale and costly program against codling moths during the early 1990s. This program was considerably more ambitious than what had previously been attempted: its objective was completely eradicating the moth rather than the more realistic and less costly objective of merely controlling this pest.

❧

The orchard industry in the Okanagan Valley during the 1980s was a community in transition, beset by intense international competition and a shrinking agricultural land base due to urbanization. The price of Macintosh and Delicious apples that made up most of the Okanagan plantings collapsed during the 1980s because of world-wide over-production of these varieties, and growers were desperately searching for some way to survive. Some tore out their old trees and planted new varieties, others turned to organic growing, and many sold out to the flood of Okanagan immigrants seeking a quiet retirement haven or fleeing Canada's cities in search of a rural lifestyle. Growers were becoming increasingly frustrated by the economic and social pressures that were transforming the Okanagan Valley and were ready for change.

The new generation of Okanagan off-farm residents also provided a receptive audience for a Sterile Insect Release program. SIR was attractive to new residents for one simple reason: they had come to the valley with an urbanite's environmental consciousness but had discovered that the reality of living in a farming community was very different from the idealized version that had attracted them to the Okanagan Valley in the first place. Their new homes were within earshot of the loud whine of air-blow sprayers and within smelling distance of harsh chemical pesticides, and they did not like it.

The selling of the SIR program to growers and Okanagan residents involved a major reality shift from the cautious conclusions of the earlier studies. The key elements were a series of economic analyses and feasibility studies, produced by Agriculture Canada for the British Columbia Fruit Growers Association, that were designed to provide convincing evidence that Sterile Insect Release was economically feasible. The optimism in these reports was in marked contrast with the cautious and limited tone of the earlier publications by Proverbs and his colleagues. These studies were almost guaranteed to modify the earlier conclusions because they changed three basic assumptions that transformed the cost/benefit analyses from clearly negative to strongly positive.

First, the new reports assumed that the codling moth could be completely eradicated from the Okanagan Valley and the valley subsequently maintained in a pest-free status. This assumption was important because it predicted that the program could shut down once it succeeded, reducing costs dramatically. Second, the new reports adopted the lowest and most optimistic estimate of how many sterile moths needed to be reared and released to achieve eradication, which also considerably reduced the proposed costs. Finally, the reports used a common accounting sleight-of-hand to amortize the costs over the longest possible term, which had the budgetary effect of making the program appear more cost-effective than it would have been using shorter-term calculations. With the changed assumptions, it was not surprising that the new analyses came up with the hoped-for result.

It is easy in hindsight to become cynical about how SIR was portrayed during this "selling" phase, but it is important to remember that everyone involved had the best intentions of reducing pesticide use and providing economically effective pest management against codling moth. What happened in the Okanagan Valley during the 1980s was that growers, politicians, scientists, and the public wanted SIR to succeed too much, and any opposition was steamrollered to the side by the enthusiasm that pervaded the SIR camp.

The atmosphere at that time was not conducive to criticism. I spoke with the director of the SIR program, Ken Bloem, in 1995 about those early days when the program was established, before he was hired in 1992. Ken is tall, thin, and casual in appearance but his relaxed demeanor belies a serious and intense perspective, and an honesty which makes it surprising that he has survived as director of such a politically sensitive program. He arrived in the Okanagan with excellent credentials, directly from the USDA Mediterranean fruit fly SIR program in Guatemala, but quickly discovered that he had moved into a program with expectations far beyond what might realistically be accomplished:

> I guess it's become clear to me, and this is just my opinion, that the comments that took place to sell the program to the regional districts, or at least the people involved from the scientific standpoint, said things that weren't fully true and people heard what they wanted to hear. They basically entered into this program from a very, very naive standpoint of what it takes to run a Sterile Insect Release program . . . I know that district horticulturists who raised some concerns have said they were basically told to shut up or stop coming to meetings. "Don't come to SIR meetings anymore because we don't want to hear your concerns. If you raise those concerns at this point in time we may not get the funding to make this program happen." It seems to me this program has always worked on the principle of "We'll do what it takes to make it get to the next step. If we build a facility and get that far along, well, maybe then we'll deal with reality." Are they really going to close this program down once they've committed seven, ten million dollars to build the facility?

Don Thomson, a technician with Agriculture Canada in the 1980s who now runs a private pest management consulting company in the state of Washington, echoed Bloem's comments: "Every entomologist was against this project, but we were told to shut up. Agriculture Canada had an agenda to be seen as bringing product to marketplace, and nothing was going to be in the way."

The steamroller picked up speed, and the combined support of the B.C. Fruit Growers' Association, Agriculture Canada, and growers of organic apples proved sufficient to obtain the approval of the five regional districts in and near the Okanagan Valley in which apples are grown commercially. With the agreement of these local governments, the Province of British Columbia passed Bill 75 in 1989, which provided the legal structure for the regional districts to administer the program and collect taxes to operate it. The approximately $8 million cost to construct the facility was to come from capital grants from the provincial and federal governments.

What was unusual about this legal structure was that two-thirds of the proposed $1.5 million annual operating costs were to come from a specific property tax in the local districts, a levy that appears on every taxpayers' bill as a line item for the Sterile Insect Release Program. The remaining costs were to be covered by growers, based on a compulsory, per-acre levy. Finally, the board overseeing the program was set up with five voting directors, one from each of the five regional districts in the eradication area, providing immediate accountability to the public for how the program operated.

The proponents of the SIR program were able to convince the fiscally conservative regional districts to proceed with this program because they emphasized eradication rather than control. This emphasis on eradication was contrary to the way most professionals practice pest management today. Eradication of pests is generally viewed as biologically difficult and exorbitantly expensive. Rather, contemporary pestologists usually stress pest "management" to maintain pest populations below economically damaging thresholds, rather than eliminating them entirely.

This emphasis on eradication may have been unrealistic, but it was not surprising. Eradication has always been the objective of Sterile Insect Release programs because of the extremely high costs associated with this method of pest control. The high up-front costs appeared justifiable to the taxpayers because they were told that the property tax would be reduced almost to zero at the end of six years and would terminate completely at the end of eight years. Also, growers were supposed to participate in funding the program, with their funding increasing toward the end of the SIR period to cover the remaining costs.

The SIR program that finally was implemented was divided into two areas, a southern and a northern region, with insect release beginning in the southern zone and moving north when eradication was accomplished in the south. The program included a two-year prerelease period, in which moth damage was to be reduced to below the 0.5 percent economic threshold by heavy pesticide spraying and removal of abandoned orchards and trees. This prerelease program was important to the SIR concept, because initial reduction in moth numbers was critical to achieve the 40:1 ratio of sterile males to fertile females necessary to achieve eradication. The final component of the program was the release of 4,800 sterile moths per acre per week, a number that was supposed to accomplish eradication in three years.

The nerve center and heart of the program today is centered inside the SIR rearing facility, located in a large gray metal-corrugated building in an industrial park in the southern Okanagan Valley that opened in 1993. By any standards, the facility is state-of-the-art, and when fully operational can churn out 10 million moths per week, twice the number for which it was designed. Everything about the facility is large-scale. Fork lifts move tons of palletized diet components to industrial-sized grain hoppers and mixers that process the 26 diet ingredients to feed the moth larvae. The developing insects are kept in sixteen self-contained, environmentally controlled rearing rooms, and the emerging adult moths are attracted by ultraviolet light to a vacuum outlet that sucks them into collecting boxes. The moths are then chilled, put in dishes, irradiated, and distributed throughout the release zone by a fleet of trucks and all-terrain vehicles.

The SIR program is now entering its fifth year of operation, and it is becoming increasingly clear that the implementation of Sterile Insect Release against the codling moth is not proceeding quite the way it was planned. I spent an evening discussing the SIR program with Linda Edwards, a resident of the valley who lives the rural lifestyle that attracts so many tourists and transplanted urbanites. Linda operates a private pest management consulting company called Integrated Crop Management, Inc., that provides advice to growers concerning the wide diversity of pests that afflict the orchard industry. She is unusual in that she moves easily between the conventional and organic grower communities and maintains the respect of both of them. Her home on an orchard in the Similkameen region is a picture-perfect Okanagan farmhouse, the walls lined with weathered barn boards, heat provided by a large wood-burning furnace, and a huge picture window framing acres of organically grown apples. Edwards is basically a farm girl from the prairies transplanted to the orchards of British Columbia, and she comfortably maintains the pragmatic values of a prairie farmer tempered by a willingness to innovate. She, like everyone else involved in the orchard industry, would dearly love for SIR to succeed:

> All of us want SIR to work and have often consequently substituted optimism for scientific scrutiny . . . We really prefer not to spray. Until you've actually gotten sick from Guthion you can't appreciate how bad that is. There isn't a single farmer who wouldn't go out and pour diesel over his sprayer and set fire to it if he thought he'd never need it again. We as growers hate filling up our tanks and going out spraying and wanting to get off our tractors and throw up. That is the single reason that growers are supporting this program, because they are directly affected. That's why we supported SIR.

Edwards continues to support the concept of SIR, but she, like other growers, extension agents, outside consultants, and many of the regional district politicians, are starting to wonder whether this program will ever work. "If it worked, it would be wonderful. For me, I run aground on reality. We wish these little suckers would work, but they don't. If I told a grower what to do based on what I wished to be true rather than what is true, I'd be out of business in a year. A lot of people in the program are removed from that reality . . . I can't pretend something is happening when it's not. People say I'm negative, but I'm just calling it as it is. If this program worked, we would be happy to pay for it, even though it's expensive. The problem with it is it doesn't work."

The problems besetting SIR have biological, economic, and political components, but in the end the program's difficulties are rooted in one issue: SIR was oversold to the public as an eradication program, and codling moth eradication in the Okanagan Valley is simply not feasible, at least with the level of resources now committed to the project.

The first problem is a failure to achieve the 40:1 ratio of sterile to fertile insects that is necessary to eliminate the codling moth. Even after cleaning up most of the wild moths in the spring with prerelease Guthion sprays and removal of infested fruit and trees, the program has had difficulty in achieving rations above 5:1 before July, and it has exceeded the desired 40:1 for only a few weeks at a few sites anywhere in the release zone.

In hindsight, the 1970s trial program in the Similkameen Valley was conducted at the wrong location. The Similkameen is a textbook case of where a Sterile Insect Release program should be successful, but it is not typical of the rest of the Okanagan Valley. The Similkameen has the most extensive area of commercial orchards left in the Okanagan Valley, so that compliance with the prerelease moth reduction program has been high and relatively easy to monitor. In contrast, the rest of the valley is a patchwork of small commercial orchards, poorly tended hobby farms, and abandoned orchards whose owners pay little or no attention to spring codling moth control. Although the law requires that owners spray every single tree in the spring or else remove it, compliance has been patchy among nonprofessional growers. Thus, the abandoned orchards and backyard plantings have served as seed sites for recolonization in commercial orchards by codling moths. In this environment, even heavy Guthion spraying by orchardists has not been successful at suppressing early spring populations to a sufficient level to conduct a Sterile Insect Release program.

Moth quality has been another problem plaguing the program. Linda Edwards described the spring-released sterile moths: "The moths were released and they would sit on the ground . . . when they land, they land on their backs." The moth problem is most acute in the spring, when temperatures are cool, often close to freezing. Ken Bloem told me: "There is some concern that the moths being reared in the facility are not as well adapted as are the wild moths to cool spring conditions. Despite our high release numbers, sterile moth counts were very low throughout May and early June. Only when outside temperature began to warm up did sterile moth activity increase."

The program's insect handlers believe that the cool weather flight problem is due to the moths not going through a cold cycle during rearing. They hope that operating the rearing rooms on a fluctuating temperature cycle that imitates cool evenings and warm days will solve this problem. Unless the sterile moths can be induced to fly at the same cool temperatures as fertile wild moths, the SIR program will have little chance of success.

❧

Growers are becoming increasingly concerned about the failure of SIR to reduce codling moth populations by the fifth year of the program, let alone eradicate the moths. Even Wayne Still, an organic apple grower who has been one of the most vocal proponents of Sterile Insect Release, is worried: "We have to have some assurance it will do what we want it to do. So far, we don't have that assurance. I'm not sure how we can sell the program to growers without a demonstrated effect that it will decrease codling moth." Ken Bloem shares this frustration at the continued high level of wild moths: "People thought it was going to be a lot easier. Growers are still concerned. One said they were told that once they started releasing sterile moths they wouldn't have to spray anymore and now they're being told they have to spray *and* release sterile moths and they just can't figure out why."

The tax-paying public also is concerned, because the combined problems of prerelease sanitation and moth viability are forcing the SIR program to reinvent the economics of codling moth eradication once again. The new numbers continue to rise and are beginning to frighten the politically responsible SIR Board. The problem for the board, according to Fred Peters, an orchardist who now works for the SIR program, was that "they got sold a bill of goods in terms of the time and cost, which both doubled on them once the program started."

The cost over-runs are creating tension among the elected politicians who must make the final budgetary decisions about this program. For one thing, the rearing facility has proven more expensive to operate than expected due to high maintenance and repair bills, breakdowns of the all-terrain vehicles that are supposed to deliver moths into orchards, and the need to hire additional personnel to do such mundane tasks as washing rearing trays.

Another source of extra expense has been public relations. The program's initial lack of success has forced the board to hire "spin doctors" to maintain public and grower enthusiasm. The budget now contains a new $100,000 line item for a private consultant to "assist the Board and Program Management in dealing with the public." Even communications with growers are not simple; there are over 900 growers in the southern zone alone, many of them recent East Indian or Portuguese immigrants who speak and read little English.

A third source of financial concern has been the cost of enforcing compliance. For the noncommercial sector, this has meant not only advertising but a major jump in staffing for personnel to take out abandoned trees and monitor urban compliance with spray requirements and destruction of infested fruit. This item does not appear anywhere in the original budget but required a $91,000 expenditure in 1995.

Even growers are beginning to balk. They are being asked to put on five sprays of Guthion per year as part of the presanitation program, which is two or three more sprays than most of them used *before* SIR. Some growers are simply not complying, while others who have complied have not been successful at reducing codling moth damage to the 0.5 percent level required in the presanitation program. All of the growers are wondering why their pesticide use has increased since Sterile Insect Release began.

The SIR program addressed this problem in 1995 by refunding the assessment of $65 per acre of orchard to those growers who met the presanitation control expectations. The effect of this rebate policy was that the 1995 program was almost fully funded by property owners, who are becoming perturbed by what appears to be a lack of financial commitment on the part of orchardists.

In total, the operating budget has crept up to $2.6 million per year from the $1.5 million originally projected, but even that increase is not the major economic problem facing Sterile Insect Release today. What is of even greater concern is that the program has yet to leave the southern zone and begin releasing moths in the northern zone. Obviously, any expansion into the north will require an additional outlay of funds, possibly doubling the already bloated budget. Even worse, the north will be a more difficult environment in which to accomplish eradication than the south, because the northern zone is more urbanized and contains smaller commercial holdings and many more abandoned trees than the southern zone.

For the SIR Board of Directors, however, one other statistic makes it clear why the program is in deep trouble: 53 percent of the funding for SIR comes from the city of Kelowna in the northern zone, the largest city in the Valley, which has yet to see the release of a single sterile moth. Property owners and the board are becoming restless, caught between the rock of a dysfunctional program and the political hard place of canceling an initiative in which $16 million has already been invested.

The board has done what government agencies do when faced with a seemingly unresolvable dilemma: buy time by hiring yet another outside consultant. The January 1996 report of the Vancouver-based ARA Consulting Group suggested solutions to the SIR dilemma that are biologically more realistic but may be economic and political nonstarters. Most significantly, the report suggested changing the focus from eradication to control: "While it would be desirable for the insect to be completely eliminated from the region, the feasibility of obtaining this goal is questionable, and the benefit–cost ratio likely unfavorable."

The report goes on to recommend redefining the concept of eradication as "the elimination of this insect species as a pest of *commercial* apple production" (emphasis mine). This subtle semantic shift would allow SIR proponents to continue using the key term "eradication," while recognizing that in reality the codling moth will not disappear from the Okanagan Valley. Rather, the moths will persist outside of orchards, acting as a permanent insect reservoir that will reinfest commercial plantings. Further, the ARA report concluded that even this reduced objective will cost more, take longer, and require deletion of some areas from the program if it is to have any chance of success.

The problem with this revised concept is that the public and the regional district politicians may not buy it. They were led to believe that total eradication of codling moth was possible, and that both SIR and pesticide use against the codling moth would soon be history. Taxpayers are unlikely to accept a $2.5 million annual bill for codling moth control if there is no end to the program, just as they would not have accepted an ongoing program when SIR was first promoted in the 1980s. Yet SIR cannot be economically viable without this public funding, since growers cannot afford the steep bill for the program on their own. There is some possibility that a provincial or federal agency could bail it out, but Canadian governments today are in a deficit-reducing mode and are unlikely to look favorably on new expenditures.

The most positive aspect of the SIR program has been the willingness of local taxpayers to make a financial commitment to an environmentally progressive program of pest control, even if it was based on overly optimistic projections of success. In that sense, the decision to proceed with SIR was a landmark in community involvement, which might have become a classic case study in how farmers and citizens could unite in a functional partnership to improve human and environmental health.

The most unfortunate aspect of the codling moth SIR program is that the growers and scientists who favored the project began to believe it could go farther than scientific studies suggested. They came to believe in eradication rather than management, although the scientific evidence to support the feasibility of eradication was weak. The perception that the public would fund the project only if it led to the disappearance of the moth from the Okanagan Valley led SIR proponents beyond the available data.

The use of expert scientific opinion in setting public pest management policy would seem highly desirable, yet even supposedly objective scientists can lose their critical acumen when they come to truly believe in the ability of science to overcome nature. The most obvious danger signal was when critics of the program were told to back off. One past and one current Agriculture Canada employee, as well as the current SIR Director, independently told me of instances when they or other critics were told to "shut up" by administrative personnel.

In the end, growers, the public, and the scientific community allowed themselves to be misled by their strong desire to do environmental good. Their objective, reduced pesticide use, could still be accomplished with the help of a Sterile Insect Release Program. But everyone involved has to realize its limitations. If SIR is to be a viable tool in managing codling moths, on-going funding will be required. SIR is not a magic bullet, but it can be part of an integrated program designed to reduce, but perhaps not eliminate, pesticide use against the codling moth.

If, on the other hand, the codling moth SIR project is cancelled, its legacy will go beyond the failure of a single program. The next time anyone in the Okanagan Valley proposes an alternative to pesticides, the collective public memory will dredge up the failure of SIR, and it will be a long time before Okanagan residents will again be receptive to a locally funded pest management program, no matter how successful and environmentally friendly it may promise to be.

CHAPTER SIX

Nature's Perfume

"A truly extraordinary variety of alternatives to the chemical control of insects is available. All have this in common: they are *biological* solutions, based on understanding of the living organisms they seek to control . . . Some of the most interesting of the recent work is concerned with ways of forging weapons from the insect's own life processes."

Rachel Carson, *Silent Spring* (1962)

Pest management is a business. It has science behind it, sometimes fascinating science, but in the end the bottom line determines whether a piece of interesting biological research is relegated to the textbooks or becomes a commercially useful system to control a pest. The farmers, exterminators, extension agents, and agricultural product distributors who make decisions concerning pest control are not swayed by elegant science, clever techniques, or trendy new ideas. Rather, decisions are made by the simplest and most pragmatic of criteria: which product or method does the best job of controlling a pest with the minimal cost.

Pest management today is still pesticide-heavy because chemical pesticides are the most efficient and direct way of meeting the joint standards of high efficacy at low cost. There have been innumerable scientific advances in developing alternative, biologically based, environmentally friendly solutions to pest management in our century, but none of them has even approached the commercial success of pesticides. One of these alternatives which has spawned enormous research interest and numerous small companies is pheromones—substances released by one animal that cause a specific reaction on reception by another individual of the same species. Yet the innovative chemical ecology industry has yet to make the major commercial breakthroughs that would launch products based on this biological method as serious competitors to the synthetic pesticide industry.

The existence of pheromones and their potential for managing insects and other pests have been known for some time. The French entomologist J. H. Fabre was the first to formally investigate the ability of insects to find one another over long distances using pheromones. In a series of classic experiments, he put female moths in wire cages on his window sill and then observed that tens or even hundreds of males were attracted to the cages. When marked males were released as far away as seven miles from the caged females, many of them appeared back at the window sill within hours. But if the males' antennae were removed or painted over with lacquer, they lost their ability to find the cages with the imprisoned females, even at short distances. Fabre speculated that male insects, especially moths, could orient to scents released by females, and that one day "science, instructed by the insect, would give us a radiograph sensitive to odors, and this artificial nose will open up a new world of marvels."

The isolation, chemical identification, and synthesis of a pheromone did not occur until the late 1950s, when the use of Fabre's "artificial nose," the gas chromatograph, was perfected by chemists. This instrument separates compounds and allows them to move at different rates through a column, where they can be detected and identified. German scientists used this new technique to elucidate the sex pheromone produced by the female silkworm moth *Bombyx mori*. This insect was an unlikely candidate to initiate the field of insect chemical ecology, because it is not a pest but a beneficial insect, and there was no compelling economic reason to find this particular pheromone. However, this moth had a number of advantages over pest species that might have been chosen for the singular honor of being the first insect to divulge the identity of its aphrodisiac chemicals. It is a large insect, and in the 1950s the technology to identify minute quantities of insect-produced chemicals was in an early and crude state. Also, methods were available from the silk industry to rear large numbers of these moths.

Even with its large size, the task of isolating enough moth-produced chemical to identify was daunting. The mating ritual of the male and female moths was well known and provided a good bioassay to test potential pheromonal compounds. The female sits on a tree trunk, everting a gland in her abdomen and releasing the attractive pheromone. The male flies upwind, using his large antennae to smell the female's species-specific scent and orient to her. Subsequent studies showed that the male antennae could respond to as little as one molecule of pheromone, and orient to the females with only a few hundred molecules released in her odor plume.

Unfortunately for science, each female produces only about one millionth of a gram of pheromone from her abdominal gland, enough to potentially attract up to a billion males but far below the detection capabilities of 1950s technology. The Munich scientists, led by A. J. Butenandt, had to clip 500,000 female abdomens to extract enough pheromone to identify its chemical structure, but they finally succeeded in 1959. They named the attractant odor bombykol, after the moth's scientific name *Bombyx,* and found that a synthetic version of the pheromone placed on a lure attracted males in a fashion very similar to a live female moth.

The potential impact of identifying the silkworm sex pheromone was not lost on the scientific and pest management communities. The 1960s saw a trickle of new pheromones isolated, identified, synthesized, and then tested as management tools to overcome insect pests with their own compounds. However, the trickle grew to a torrent as techniques improved, instrumentation became more sensitive, and basic knowledge concerning pheromone-based biology created an increasingly sophisticated substrate on which subsequent researchers could build. The growing commercial interest in pheromones, and our increasing technical capability to identify them, was reflected in the number of U.S. patents granted for novel pheromones. There were only 13 patents granted before 1970, but 150 were granted by 1988, and 257 by 1991.

Interest in pheromones has expanded into the discipline of chemical ecology, which includes not only pheromones but any chemical involved in communication between organisms. A new term, semiochemical, from the Greek *semion,* to sign or signal, was coined to reflect the broadened scope of scientific inquiries into chemical communication. Today's chemical ecologist might still elucidate the identity and function of an insect sex pheromone but is just as likely to study the odors that attract a pine beetle to its host tree, the inhibitory secretions that prevent a worker bee from laying eggs, or the alarm chemicals given off by an aphid that is under attack by parasitic wasps.

The work of these chemical explorers has been of considerable interest to pest managers, because semiochemical-based pest management has great potential advantages over more traditional pesticide-based control. Most significantly, pheromones are highly specific to individual species. Although it is not unusual for related insect species to use one or more of the same compounds as pheromones, the blend of chemicals produced is unique to each species. The blend of sex pheromone that attracts the male of one species will be ignored by other species, providing a specificity to pheromonal-based management that is lacking for pesticides. Thus, pest management using pheromones has no impact on nontarget organisms, a tremendous advantage over more broad-spectrum chemical pesticides.

Pheromones also are highly active at sometimes unbelievably low concentrations, with many insects responding to only a few molecules. The industrial production for the world-wide use of any pheromone for pest management runs to only a few pounds each year, whereas pesticides are produced by the tons. Although pheromone syntheses can be complex and costly, the infrastructure required to produce commercial levels of pheromones can usually be fit into a small laboratory rather than the industrial-sized plants needed for pesticide manufacture.

Pheromones also have the advantage of being relatively easy to register and market, because they have virtually no side effects on vertebrates or even other insects. Their nontoxic nature is due to two factors. First, the type of chemical structures found in pheromones tend to be benign toward most organisms. Second, and probably more significantly, the quantities of pheromone set out for a control program are ridiculously low from a toxicity perspective, so that the impact of even the most potentially toxic pheromone would barely register on most organisms.

The practical uses of pheromones in pest management have settled out into three main techniques: monitoring, attract-and-kill, and mating disruption. Monitoring is the most common application of pheromones to pest management. Typically, an open-sided trap is set out with a lure inside that is baited with the target insect's sex pheromone. The attracted insect enters the trap expecting to find an insect of the opposite sex, but instead encounters a sticky lining from which it cannot escape. Pest managers check these traps on a regular basis, correlate the numbers of trapped insects with potential economic damage, and make informed decisions about when and how often to apply pesticide sprays. The use of pheromone traps requires some expertise in interpreting the results but, when properly used, can reduce the number of chemical sprays.

The attract-and-kill technique is particularly effective against insects in enclosed spaces, such as beetles in grain bins or cockroaches in interior urban settings. The approach is similar to monitoring in that a trap is baited with the attractant pheromone, but the insect that enters the trap is met with a contact poison and quickly dies. The advantage of this technique is that insecticides can be contained within traps and do not enter the environment.

In the third commonly used strategy, mating disruption, the air is permeated with sufficient quantities of synthetic pheromone to confuse insects attempting to locate potential mates. Pheromone-releasing dispensers are placed at regular intervals in fields or forests during mating season, and most of the confused insects fail to mate, thereby reducing the next generation's population. This method is difficult to implement in practice because of problems in releasing sufficient pheromone to saturate the airspace above a field on a continuous basis and at an economically affordable price. Mating disruption also has no impact on the damage caused by the current insect generation. Nevertheless, it is probably the most commercially profitable method in those few systems in which it is being used today.

Pest management using semiochemicals has had some success, and there is no question that chemical ecology has the potential to provide outstanding tools to control pests with minimal environmental impact. Further, the possibility of using pheromones to manage pests has fascinated the public, and reports about pheromones are prominent in the media. Semiochemical research at its best provides hope that human ingenuity can out-smart pests without synthetic pesticides. However, reports of pheromones in the news generally describe the initial research breakthrough and its potential for commercial application. Few of these media-reported pheromone stories actually make it to commercial viability.

Pheromone-based studies have matured and flourished as one of the major niches of contemporary biology, but pheromones as an industry have not yet developed as more than an interesting sideline to the mainstream pesticide industries. The pheromone industry today consists of many small, short-lived companies with marginally successful products; only a few companies or products achieve substantial success. Even the most outstanding pheromone-based products required extensive financial support by government before they reached a commercialized stage.

Statistics compiled by Mike Banfield, co-founder of a small pheromone company called Phero Tech, reveal a small, struggling, and highly volatile semiochemical industry. Total 1991 sales were only $38 million from 17 North American firms that synthesized and formulated 139 different pheromone products. About one-third of those sales involved pheromones used in cotton pest management to monitor and control boll weevil and pink bollworm pests, with annual sales for these two products reaching $6 million and $7 million respectively. The most significant customers for pheromone products are agricultural supply companies, which then distribute them to farmers, and government agencies that run large pest management programs on public land or for research purposes. Mating disruption and attract-and-kill pheromones made up $10 million in sales, with the remaining products formulated for monitoring.

All of these small companies are hoping for a breakthrough, an easy-to-produce, patentable pheromone management system that can earn big money. Few companies find the pot of gold, however; most pheromone products are low in sales volume and are produced by a number of competing companies. The average sales for any single pheromone product were only $275,000 per year in 1991, with individual firms producing an average of three to four products each. Although some commercially available pheromones are sold by one company, most products have up to eight companies competing for the same market.

The two most significant products, boll weevil and pink bollworm pheromones, are sold by six different companies, giving each company an average market share of only $1 million for each insect's pheromone.

Codling moth pheromones exemplify the effort required to bring a semiochemical product to the marketplace, and the difficulties in generating enough income to make the business end of pheromone-based management profitable. They also provide an interesting contrast to the publicly funded SIR project in the Okanagan Valley of British Columbia. Today, an area-wide program is under way to disrupt codling moth mating with pheromones in Washington, Oregon, and California, just south of Canada's Sterile Insect Release program. The early results have been positive enough to stimulate a rise in sales of pheromone products for use in codling moth pest management, but this mating disruption program also illustrates why pheromones remain a minor component of pest management, even with effective products.

❧

The codling moth program in place today has taken twenty years to implement and has been possible only because of the combined efforts of university and government researchers, extension agents, private enterprise, and growers. The development of pheromone-based codling moth management began with the chemical identification of the primary component of the codling moth sex attractant, codlemone, in the 1970s by Wendell Roelofs and colleagues at Cornell University. That, however, was only the first of many steps leading to a commercially viable product.

The first barrier to overcome was that natural extracts from female codling moth glands were 1,200 times more effective in attracting male moths than equivalent amounts of synthetic codlemone. This is a common problem in pheromone research but also an intriguing challenge for chemical ecologists, because almost all insect pheromones are multicomponent and can include five or more compounds.

Unfortunately, the least abundant and most difficult to identify compounds in the blends often are crucial to synergize the chemically most abundant components. It can take many years of challenging chemical detective work to identify these minor components. Chemists must first test crude fractions of gland extracts in all possible combinations to determine where active compounds are found in the extracts. Then, each compound in active fractions has to be identified, synthesized, and tested in laboratory bioassays in combination with other potentially active compounds. Finally, promising chemicals need to be field-tested to determine whether they do indeed synergize the known major component. For codling moth, it took ten years from the initial identification of codlemone to find two other components, dodecanol and tetradecanol, that were required to improve the efficacy of codlemone into a commercially useful range. Even today, more than twenty years after the identification of codlemone, this three-component blend is not as active as the female-produced blend, and scientists continue to search for the remaining components.

A second problem, and by no means a trivial one, concerns how to release codling moth pheromone in the field to confuse male moths sufficiently to disrupt mating. This problem is compounded for codling moth because the major component in codlemone is sensitive to light and quickly breaks down to inactive compounds under direct sunlight. Further, the pheromone blend has to be released at a consistent dose every evening when the moths fly for the three-month period of moth flight; if the dose is too low, the male moths will not be prevented from finding the females and mating. Pheromone release rates that are higher than necessary can be prohibitively expensive because of the significant costs involved in pheromone synthesis.

The first ten years following codlemone identification were spent finding other components, but much of the subsequent ten years has been spent testing release devices in the laboratory and then in the field. Four different devices have been developed by private companies that seem to meet the dose and release requirements; the two major products are Checkmate CM, produced by Consep Membranes, and Isomate C+, sold in the Northwest by Pacific Biocontrol, a division of Shin-Etsu Chemicals, a Japanese company. The Checkmate device is a flattened membranous polymer that allows the pheromone to diffuse at the desired rate through a plastic laminate that protects it, while the Isomate C+ dispenser is a brown polyethylene tube resembling a twist tie that has pheromone imbedded in it; the pheromone diffuses through the semipermeable walls of the dispensers. Isomate C+, the more successful product, is placed in orchards in the spring by wrapping about 400 dispensers per acre evenly distributed on branches at a height two to three feet below tree tops.

Determining how to apply these devices has not been a simple task, because of the large number of dispenser characteristics that have to be experimentally validated and the extensive seasonal and site variation inherent in agricultural research. A simple question like "At what height should Isomate C+ dispensers be placed?" can take years to resolve. A typical experiment might involve ten sites, with each site anywhere from one to ten acres in size, and experimental treatments including three or more different heights to be tested. Further, this experiment must be conducted for at least two or three years to account for seasonal variation in temperature, wind speed, and codling moth numbers. While all this is going on, growers must be compensated if an experimental treatment fails to control codling moth, since no grower can afford to let a failed experiment devastate his income. Putting all this together, it can cost millions of dollars just to determine the right height to place a pheromone dispenser. Then there are still many other factors to be examined such as dose, release rates, and orchard characteristics such as slope and size.

After twenty years of pheromone identification and dispenser testing, the use of codling moth pheromones to disrupt mating still is not economical, both because pheromone application is more expensive than pesticides and also because mating disruption is not effective in some situations. Don Thomson, who currently heads Pacific Biocontrol, told me that the cost of an Isomate C+ program when adjusted for material, labor, and machinery costs, is $85 per acre higher than a conventional insecticide program. Similarly, William Quarles, in *Integrated Pest Management Practitioner* magazine, cited direct pheromone costs of $125–215 per acre, compared with more conventional insecticide costs of $30–75 per acre.

In addition, mating disruption is not always effective. This technique requires low initial moth populations, flat orchards, and the simultaneous treatment of large acreages to work well. It is difficult to maintain pheromone saturation in orchards on slopes or in windy locales, and mating disruption does not fully prevent matings in orchards with high moth infestations because enough moths encounter one another to overcome the pheromone's effects. Even under ideal conditions, mating disruption is not as effective at orchard edges as it is in the more central sections.

Nevertheless, in spite of high up-front development expenses, higher cost to growers compared with pesticides, and limited situations in which it is effective, companies like Pacific Biocontrol have been able to sell sufficient product to justify their commercial involvement. An important factor in the commercialization of codling moth pheromones has been the intervention of the federal government, which kick-started the use of pheromones on an area-wide basis and continues to subsidize research and development costs heavily. Growers are highly averse to unnecessary risks; any new technology must be clearly demonstrated to be effective before farmers will risk their crop. Further, the research and orchard-level demonstrations necessary to convince growers that an alternative method such as mating disruption will work are expensive, much more costly than the research budget of a small pheromone-producing company can even begin to afford. Thus, stimulation provided by government funding has been necessary to bridge the gap between concept and application.

University and government researchers, extension agents, and orchardists belonging to the various fruit growing associations in the Pacific Northwest put together a proposal for the U.S. Department of Agriculture to consider in the early 1990s. At that time the USDA was discussing a shift in funding from individual projects toward consortium-based, area-wide projects focused on particular pest problems. The codling moth mating disruption technique was a perfect candidate for this new program, and the government provided about $1 million a year for three years to develop this technology for private sector use.

The area-wide program for codling moth management involves 93 orchards covering 2,000 acres at five different sites in the Pacific Northwest. This program has demonstrated that mating disruption can achieve damage levels similar to those found in pesticide-based programs—equal to or lower than the 0.5 percent threshold for fruit damage. Consequently, the acreage in which mating disruption is being used by grow-

ers has risen from 1,500 acres in 1991 to 18,000 acres in 1995. Although this represents less than 4 percent of the total apple acreage in the United States, it at least shows growing confidence in the product.

The proponents of mating disruption have greatly enhanced its credibility as a pest management technique against codling moth by the realistic approach they have taken. Don Thomson of Pacific Biocontrol and Jay Brunner from Washington State University, a leading researcher in orchard pest management, both have been involved in the area-wide project for codling moth management. I spoke with them in the unusual venue of a Las Vegas casino, where the Entomological Society of America was holding its annual meeting. They both had the same carefully considered, realistic approach to the benefits and disadvantages of mating disruption—a sharp contrast to the ring of slot machines and high-risk gambling fervor that surrounded us.

Brunner considers mating disruption to be only one of many management paradigms for codling moth control. His attitude is that pesticides, too, are an important component of orchard pest management but should be the method of last resort: "Integrated Pest Management has been Integrated *Pesticide* Management. If you think about the concepts behind ecologically based pest management, you would use these powerful, useful chemical tools last, not first. I would like to see us conserve these tools for future use, because we'll need them from time to time, but not as first choices. Mating disruption has allowed us to do that." Brunner recognizes that mating disruption will not work in every situation, however, and is not averse to advising growers to spray pesticides against codling moth when pheromones fail to do the job.

Both Brunner and Thomson also are critical of the eradication approach taken by the Sterile Insect Release Program in the Okanagan Valley to the north of them. A key element in the success of the pheromone project compared with SIR has been its focus on moth management rather than elimination. Brunner is against legislated pest management, believing that growers need to be convinced to use an alternative control method rather than be forced by heavy-handed legislation to adopt an eradication approach: "We're looking at suppression as a strategy rather than eradication. It's a completely different mindset. Eradication takes tremendous energy to achieve, if it's even possible, and requires legislation of a program to everybody. I'm completely opposed to legislating pest management." Thomson agrees, especially where the legislated SIR program is concerned: "The SIR program is flawed because its objective is eradication. Management is the real approach. SIR can never be successful because eradication is unreasonable."

Perhaps the most significant factor in the commercialization of mating disruption productions has been the decision to market products like Isomate C+ as a broadly based pheromone management system rather than just as an isolated product. This approach is important, because it would not be possible to sell codling moth pheromone based on direct costs and benefits, since insecticides come out significantly cheaper. Thomson's sales approach has focused on two major advantages to mating disruption that are more indirect yet important to growers. First, pheromone management is environmentally friendly, and growers are as interested in reducing pesticide use as anyone else, even if costs are somewhat higher. This interest is not completely "green"; growers recognize that codling moth has a long history of developing resistance to pesticides and have been predisposed by recurring resistance episodes to try alternatives.

Another significant economic argument used to persuade growers to try mating disruption has been that this technique reduces the need to use so much pesticide against the many other pests that infest orchards. For example, the cost of insecticides for other pests was $90 to $150 lower in orchards where codling moth was controlled with mating disruption, as compared with orchards where Guthion was used. The main reason for the reduced pesticide sprays and costs in the pheromone-treated orchards was the increased natural populations of predators and parasites that controlled other pests such as mites, leafrollers, aphids, and leafminers; beneficial pest-attacking insects are killed off by synthetic chemical pesticides, and their absence allows apple pests other than the codling moth to proliferate.

In addition, mating disruption leaves no pesticide residues on fruit and does not prevent workers from entering fields during and after sprays. These are not economically trivial or irrelevant considerations to growers. When the costs of resistance, increased pesticide use against other pests, residue analyses, and down-time for workers following spraying are considered, Thomson calculated the indirect price of spraying Guthion on apples against codling moth at an additional $182 per acre beyond the direct costs of purchasing and applying it.

Thomson's marketing approach has been to stress these indirect costs to balance the higher direct expense of pheromone compared with Guthion. He recognizes that "even though long-term costs of mating disruption may be cheaper, the short-term, up-front costs in conjunction with uncertainty of control has kept a lot of growers from investing in this technology." At least some growers are accepting his more indirect and long-term cost–benefit analyses. Isomate C+ sales for mating disruption rose from $350,000 in 1991 to approximately $1.8 million in 1995.

Although the codling moth mating disruption program has slowly approached commercial viability, it nevertheless remains an economically minor component of apple pest management compared with pesticide use, even with the most optimistic industry sales projections. It has taken great effort to reach this point, and it is discouraging to advocates of biologically based pest management that sales of codling moth pheromones remain almost trivial relative to hard chemicals.

The same is true for virtually all pheromone-based management. The $38 million in annual sales of all pheromones and related products in North America is less than a tenth of 1 percent of the $8.5 billion in annual sales racked up by pesticides. Despite high levels of government financial support, great interest and activity by the research community, and years of product development, pheromones remain in the world of alternative rather than mainstream technologies.

Given the problems of commercially developing codling moth pheromones, it is apparent why pheromones and other alternative technologies are still fringe players in the pest management business. Certainly one factor that has limited the growth of pheromones as a practical pest management tool has been the technical difficulty in developing products. The initial identification of a pheromone is challenging enough, but there is a complex array of information still needed to reach the final product stage. As Mike Banfield put it, "The greatest contingency faced by firms in the semiochemical industry is information complexity . . . A product targeted at a single insect pest may need up to six chemical compounds, each of a different purity, quantity, and release rate. These chemicals must be enclosed in a protective device comprising a chemical reservoir and slow release mechanism. The semiochemical blend must then be affixed to a trap, be attached to a host plant which is sacrificed to the pest hordes, or applied by hand or mechanically by ground or air."

A second impediment that semiochemical firms face is the difficult market farmers represent. They are accustomed to pesticides that are simple to apply and do not require much in the way of technical advice. The timing, rate, and dose of pheromone applications, by contrast, can change dramatically year to year and site to site, so that a trained consultant is necessary to administer pheromone-based controls properly. Even the apparently simple monitoring of pest populations with pheromone traps can be difficult to interpret and requires experts to translate trap catches into control advice. For these reasons, the consultant package that is usually sold along with the production might include weekly or more frequent on-farm visits by a technical expert and detailed bulletins explaining the various contingencies that might develop in a pheromone-based program. The need for expert advice to use pheromones properly not only increases the cost to growers but appears excessive to farmers who have simplistic  pesticidal solutions available for purchase at the nearest agricultural supply house. According to Banfield, "Convincing a grower that the method he has been using for 30 years can be superseded by a product that doesn't even kill the pest can be difficult . . . Conventional pesticides provide the farm manager with a low risk approach; he knows from experience that they work, the products are backed by large firms which can afford to buy crops if losses do occur, and he already understands how to use them and has the equipment at hand."

Another barrier to the commercial success of pheromone products is the lack of a distribution system. Most pesticide companies are large enough to support their own web of distributors, or they use agricultural supply companies that distribute numerous chemical products to growers. The approach at either level is highly personal, and dependent on growers becoming familiar with the person selling them chemicals. This network takes many years and a major financial investment in personnel and travel costs to develop and has not been particularly open to the entry of novel, alternative products. In contrast, pheromone companies invest most of their resources into research and development and have little surplus available for effective marketing and sales effort.

The market profile of pheromone products and the highly competitive nature of the semiochemical industry also has worked against the rapid growth of pheromone-based control. Pheromone marketing involves numerous small sales rather than a few large ones, so that marketing expenses and overhead are relatively high on a per-sale basis. In addition, a semiochemical product is competing not only with pesticides but also with semiochemicals produced by competing firms. The difference between one company's product and another are often subtle and involve only slight modifications in technology, price, longevity, and method of application. Thus, it is difficult for a company's product to stand out and be easily differentiated from a competitor's product.

Customers' confidence in pheromones has been further challenged by the regular and rapid disappearance, merger, or buy-out of companies. Customers worry that the continued existence of any semiochemical firm or product is unreliable. For example, Albany International, which began as a diversified pulp and paper business, in the 1970s purchased the Fabric Research Corporation, which produced hollow fibers that could be used as dispensers to release pheromones slowly. This new division was then split off as ConRel, a wholly owned subsidiary, but a few months later was reabsorbed by Albany, which dropped the ConRel name. In 1983 Albany International experienced a leveraged buy-out, and all of its divisions and subsidiaries were sold off in 1984. Eventually, the remains of the pheromone section wound up in a company called United AgriProducts, which renamed the division Pest Select International. When this corporate rearrangement failed to generate revenue, the management was fired and the remaining company was renamed Scentry, Inc., and moved to Billings, Montana. Scentry in turn was bought out in the early 1990s by EcoGen, Inc., a Pennsylvania-based biotechnology company whose main product lines are genetically engineered varieties of *Bacillus thuringiensis*. In January 1996 the Monsanto Company purchased a 13 percent interest in EcoGen and agreed to fund a four-year $10-million research and development program for *B.t.* products. Industry rumors suggest that the pheromone product lines will be dropped as EcoGen attempts to reduce the $23 million loss the company suffered in 1995.

Even relatively large and stable companies with robust sales are not making profits in the semiochemical business. Shin-Etsu, for example, which owns Pacific Biocontrol, is a division of the huge Japanese conglomerate Mitsubishi. This international division, which entered the pheromone market in 1978, currently sells 41 insect pheromones and numerous release devices, accounting for sales of about $12 million in 1991. Their product lines include two of the most successful pheromone products on the market today, Isomate C+ (used to control codling moth) and PB-ROPE (used to disrupt mating in the pink bollworm). The Japanese government, like the U.S., has been heavily involved in subsidizing semiochemical research. Yet according to Don Thomson, in spite of their extensive product line, corporate longevity, government support, and seemingly strong sales, Shin-Etsu is "still economically fragile, and is not making money world-wide on pheromones."

Finally, the semiochemical industry has suffered in the past from its failure to influence the federal government to loosen restrictions on pheromone testing and use. To address this problem, in 1992 twelve semiochemical companies banded together to form a trade association called the American Semiochemicals Association (ASA), and lobbying efforts by this group quickly demonstrated the importance of a trade association to new high-technology industries. The ASA convinced the Environmental Protection Agency that pheromones as a generic class of pest control product were environmentally benign, and the EPA changed its registration requirements accordingly so that the average time and cost to register a new pheromone product were reduced from two years and $100,000 to two months and $10,000. In addition, the EPA was persuaded to increase the acreage on which pheromones could be tested from 10 to 250 acres and also to allow both food and nonfood crops from these fields to be harvested; previously, the government had required that agricultural products from field tests be destroyed. All of these exemptions from and changes to EPA regulations will make pheromone product development considerably cheaper and faster in the future.

Pheromones and other semiochemicals represent many things that our scientific and pest management communities can be proud of. It has taken enormous ingenuity and technical expertise to identify these chemicals from the myriad insects that produce them. Our increased understanding of pest biology and the clever methods we have devised to use this knowledge for pest management provide considerable hope  that the future of pest control may lie in manipulating pest behavior rather than in applying toxic chemicals. Semiochemical research also has been a shining example of how interdisciplinary cooperation can synergize entire new fields of study to make stunning breakthroughs in chemistry, biology, and pest management.

Semiochemicals have not proven to be a commercially attractive alternative to pesticides, however. The reasons are many and complex, but the bottom line is that pesticides are cheaper to buy and apply than pheromones, are easier to use, and are more consistently effective. Semiochemicals will remain a fringe player in the pest control industry until that situation changes. This alternative technology clearly has the potential to become a major component of pest management, but for now semiochemicals with few exceptions remain in the hopeful world of possible tools rather than the actual world of commercially successful products.

CHAPTER SEVEN

Bees and Other Beneficials

"Vaccination, Gas, the Steam-engine, the Steam-boat, the Rail-road, the Electric Telegraph, have all been successively the laughing-stock of the vulgar, and have all by slow degrees fought their way into general adoption. So will it be with the artificial importation of parasitic insects . . . The principle is of general application; wherever a Noxious European Insect becomes accidentally domiciled among us, we should at once import the parasites and Cannibals that prey upon it at home."

B. D. Walsh, *Practical Entomologist* (1866)

Picture a late winter scene in the southern United States, with the first hints of spring already in the air. Beekeepers are loading their hives onto large semitrailers, following a migratory imperative that has earned them much of their living for the second half of this century. Their bees spent the winter in the South, in Florida, Georgia, Texas, Arizona, or California, but now it's time to move. Each semi is loaded with hundreds of colonies and driven to the flowering almond trees of California, then on to apple orchards in Oregon and Washington, then out to the Dakotas to get a summer clover honey crop. Or perhaps they travel first to orange orchards in Florida, then up to Maine for the blueberries, and back down to the southeast Appalachian mountains for summer honey.

The business of moving bees is not designed primarily to produce honey but rather to pollinate crops. This managed pollination service is an essential component of food production; without managed bees, many crops would not exist. About 30 percent of American bee colonies are moved to crops each year, over a million colonies. Each is moved on average to two different crops, with beekeepers receiving anywhere from $10 to $50 for each colony set down near a blooming crop. The value to beekeepers of this industry is about $40 million annually, but the worth of this pollinating service to the rest of us is almost incalculable.

The best estimate for the dollar value of crops pollinated by bees is about $10 billion a year in the United States and Canada. This figure, as high as it is, does not fully account for the impact of managed pollination. Imagine a trip to the fruit, nut, and vegetable sections of your local grocery store and consider the gaps that would appear on the shelves without managed bees to pollinate crops. There would be few, if any, apples, broccoli, avocados, cherries, cucumbers, melons, carrots, oranges, pears, pumpkins, squash, blueberries, grapefruits, macadamia nuts, raspberries, plums, onions, and many more. In fact, except for corn and a few other wind-pollinated crops, there would not be a fruit and vegetable section in the supermarket were it not for pollination by bees and a few other insects.

Most fruits, nuts, and vegetables require bees to pollinate their flowers, either to produce any crop at all or to increase crop production to a profitable level. All of these crops and their evolutionary ancestors have been around for many millennia, some for tens of millions of years, but until this century these plants were not dependent on human management for pollination. Many advanced plants evolved with bees in a unique coevolutionary contract; their flowers produced nectar and pollen to attract bees, which use these commodities as food for themselves and their brood. In the process, the hairy bees move pollen from one flower to another, thereby fertilizing the flower's eggs and producing fruit and seed.

This long-standing natural contract has been broken in this century by human intervention. Bees have been around almost as long as flowers, but the wild bee is rarely seen on farms and in orchards. We have destroyed the wild bees in our heavily managed agricultural regions and have had to substitute a managed system for what nature used to do on its own.

The loss of wild bees and their replacement with managed bees to pollinate crops provide a good example of how we can create new problems for ourselves by rearranging nature through agriculture and pest management. Further, managed bees exemplify how we have chosen to solve these problems. Rather than addressing the issue directly and developing techniques to foster and enhance feral bees, we have created the novel, artificial pollination industry, an industry that is necessary only because we modified the natural world.

Today's managed pollination industry uses only two species of bees to pollinate field crops: the honey bee, which is used for most crops, and the alfalfa leaf-cutter bee, used to pollinate alfalfa for seed production. Compare this two-species service to the 5,000 native bee species present in North America, and the extent of our impact on natural pollination begins to emerge. Our fields and farms used to be full of bees, with abundant wild species filling the spring and summer air with the buzzing sounds of pollination. These feral bees came in a great variety of sizes, shapes, and lifestyles, from the enormous, hairy, social bumble bees to tiny solitary sweat bees, from bees that nest in holes in the ground to those that live in stems and twigs. These bees have exotic names like mason, carpenter, bumble, and leaf-cutter, telling us of each bee's special skills. The masons are proficient brick makers, molding mud into cells for their brood, while the carpenter bees are superb woodworkers, tunneling into trees with their huge jaws to build their nests. Leaf-cutter bees are skilled weavers, cutting leaf pieces, carrying them back to hollow stems, and lining the stems with leafy insulation to protect their young. And then there are the bumblers: ponderous and stately, they fly erratically and noisily through the air, buzzing on flowers to dislodge pollen, and carrying enormous loads of nectar and pollen back to their subterranean nests.

These bees have not vanished from the world, but they have almost disappeared from agricultural regions. The demise of feral bees is not a major environmental issue that has eco-militants lying down in front of bee hives. Rather, it is one of many issues that has crept slowly into our consciousness, without our ever having made a purposeful decision that things should be this way.

The reasons that wild bees no longer visit crops are few, and clear:

pesticides, lack of floral diversity, habitat destruction, and, ironically, competition with managed pollinators. Pesticides are probably the single most significant factor reducing wild bees, for the simple reason that feral bees cannot avoid pesticides. Managed bees can be screened from flying or moved during sprays, but feral bees do not have that option, and they get hit hard by pesticides used before, during, or after bloom.

The vast acreages of single-crop plantings that characterize agriculture today provide a second barrier to feral pollinating insects. Most bees need a number of flowering species whose blooming periods collectively spread over weeks or months to provide enough nectar and pollen for their young. Agriculture concentrates on a single floral species that blooms over a short time span, and often one whose flower produces nectar or pollen but not both. Floral diversity in cultivated regions is further diminished by herbicides that eliminate weeds, which to a bee represent prime food sources rather than pest plants.

Habitat destruction is a third factor that diminishes wild bees, which nest primarily in soil or hollow stems. Plowing ground, turning soil, and killing weeds destroys nesting sites, and also kills bees that may be waiting to emerge from previous years' nests. Also, some of our best agricultural land is increasingly close to cities, so that urban development encroaches on potential nest sites for feral bees near crops.

Finally, managed bees are so abundant that their populations suppress feral bees. The 100 billion individual honey bees in managed beehives today are certainly having an impact on feral bees by taking much of the nectar and pollen that is available.

Thus, agricultural practices, and especially pest management, create environments that make it exceedingly difficult for feral bees to conduct their business. There are almost no wild bees in cropped systems, as a few of my students documented in a four-year study in the 1980s. They compared the abundance of bees on fruit and berry crops with those present in noncultivated areas, and the results demonstrated why managed pollination has become an essential component of modern agriculture. They collected an average of only four wild bees per hour in orchard settings, compared with the hundreds of bees per hour that are needed for effective fruit pollination. In blueberries, they observed about 33 wild bees per hour, considerably fewer than the 170 bees per hour considered to be the minimum necessary for commercial blueberry pollination.

The farmer's solution to these low numbers of wild bees has not been to steward feral bees better but rather to invent managed alternatives to compensate for the dearth of native pollinators. The most well-known of these solutions is the honey bee, a bee that was not present in North America until European settlers brought colonies over from the Old World, beginning in the 1600s.

Honey bees were originally introduced to provide honey, not pollination, and they have thrived throughout most of the United States and southern Canada. Many swarms have escaped from beekeepers' hives and established feral populations, so that today the honey bee is both a feral and a managed insect. There are about 3.5 million managed honey bee colonies in the United States and Canada, clearly an immigrant success story of unparalleled insect proportions.

The introduced honey bee is considered to be a beneficial insect because it produces honey and wax and pollinates our crops, but it also has its own imported pests. The beekeeping industry was blind-sided in the mid 1980s by the accidental introductions of two mite species into North America. The first arrival was the tracheal mite, a European species that infests the breathing tubes of bees. The tracheal mites were followed shortly afterward by an external Asian mite called varroa, which feeds on brood and adult bee blood and can kill a colony within months. Ironically, these mites have forced the previously pesticide-averse beekeepers to introduce $10 million worth of miticide-impregnated strips into colonies annually to control these pest mites. Even so, beekeepers lose an average of 25 percent of their colonies each year to mites, and shortages of bee colonies for pollination are becoming a common concern for beekeepers and growers alike.

The introduction of bee mites illustrates an important aspect of pest management that has been repeated countless times in North American pest history. Importations of exotic species invariably lead to complexities in human–pest interactions that are difficult to anticipate or predict. For bees and pollination, the deliberate introduction of honey bees was a major factor in reducing populations of beneficial native bees and led to our growing dependence on managed pollinators. Dependence, in turn, has led to vulnerability, and the presence of accidentally imported bee mites now threatens the extensive managed pollination system we developed to substitute for the natural system we destroyed.

Bees are not the only feral, beneficial insects that we have driven from our fields, and managed pollination is not the only system we have developed to compensate for the destruction of beneficial organisms by human practices. Parasitic and predatory insects that used to control many pests also have vanished from our cities and farms, victims of urban and agricultural development and of pesticides that do not discriminate between friend and foe. We have had to develop new industries to substitute human-reared predators and parasites for the now-diminished natural ones, just as many beekeepers rear and manage honey bees to perform the tasks that natural, feral bees used to accomplish.

There is a general phenomenon here that most of us are unaware of and don't consider when we purchase an apple or buy a bag of nuts. The foods that fill our markets are not the products of nature but rather the end result of intensively managed systems that have replaced what nature used to do on its own. If there ever was a "balance of nature," we have eliminated it, and much of contemporary agriculture is designed to restore this balance through management, using artificial replicas of natural processes.

Bee pests are just one example of the extraordinary number of pest management problems that have been imported to North America from elsewhere. Myriad exotic pests have devastated agriculture and resulted in serious human health problems. Further, imported pests have stimulated the growth of yet another management industry, the importation and production of beneficial agents to control both indigenous and imported pests.

About 2,000 insect and mite species of foreign origin are now known to be in North America, and this number increases by an average of 11 species annually. Half of crop losses caused by insects and mites are due to these imported species. The earlier arrivals came in rubble and soil used as ship ballast, while more recent imports have come on plants, seeds, and fruits imported for agriculture or as food. Of the approximately 600 economically significant insect pests, 235 of those are of foreign origin. Examples of exotic insects read like a most-wanted list of harmful pests: codling moth, gypsy moth, fire ants, citrus whitefly, Japanese beetle, pear psylla, San Jose scale, and European corn borer, to name just a few.

Imported plant pathogens and weeds are just as problematic. About 60 percent of our weed species were introduced to North America in the last few hundred years, and 19 of our 29 worst weeds came from overseas. Weeds such as crabgrass, Japanese knotweed, water hyacinth, wild oat, Russian thistle, bindweed, chickweed, dandelion, mullein, kudzu, and plantain plague our fields and backyards, and they all originated outside of North America. Many plant diseases, including numerous rusts, blights, smuts, spots, mildews, and cankers, had a similar exotic origin.

In nature, pest populations generally are kept under control by predators, parasites, diseases, weather, and food shortages. Pest outbreaks occur because of an increase in food supply, decrease in natural enemies, unusually favorable weather conditions, or some combination of these factors that proves advantageous for the survival and reproduction of pests. The insect pests that we have imported find themselves in a new environment with abundant food provided by crops and few natural enemies; the result is a population outbreak. Further, agriculture selects quickly for species or varieties that can thrive in farm habitats. And the distribution system of agriculture ensures that a successful pest is likely to be transported to other environments where new populations can thrive.

The use of biological control against exotic pests involves searching for natural control agents in the pest's home territory and then importing, propagating, and releasing one or more of them to combat the introduced pest. This strategy appears simple and environmentally friendly, but the concept of biological control has proven more difficult to implement in practice. Biological control also poses a major challenge in our pesticide-ridden agricultural systems, since biological control agents often are highly susceptible to pesticidal chemicals.

Biological control has been useful in a number of situations; in fact, the first serious North American attempt at using this technique was so wildly successful that it raised false expectations that biological control might be triumphant against all pests. This initial nineteenth-century experiment was designed to combat an imported pest insect named the cottony-cushion scale, a minute, flat insect that is covered with waxy protective threads and feeds on citrus, pear, and other crops. This scale insect was first found in California in 1868, feeding on acacia plantings in a horticultural nursery. It soon became established in southern California and decimated the developing citrus industry, to the point that growers were burning or digging up their trees and abandoning their orchards.

At this point Charles Valentine Riley entered the picture, the same C. V. Riley who was instrumental in initiating pesticide sprays against the gypsy moth in Massachusetts and who also patented the first successful pesticide spray nozzle. Riley had been state entomologist in Missouri but was promoted to Chief Entomologist for the U.S. Department of Agriculture (USDA) Division of Entomology about the time that cottony-cushion scale was becoming a problem. He was a brilliant entomologist, but he also knew how to work within the government system to maximize his effectiveness. His reputation was based as much on the detail, illustrations, and high production quality of his annual state reports as on his skills as an entomologist, and his reports had earned him both promotions and a devoted following among American entomologists.

Riley proposed that the cottony-cushion scale must have come from the South Pacific, since much of the citrus nursery stock used in California had been imported from that region. Also, the insect had been described by a New Zealand entomologist as being native to Australia. Riley discovered through correspondence that the scale insect was a serious pest in New Zealand but not in Australia, and then made one of those tremendous jumps in reasoning that proved to be a key moment in pest management history. He surmised that cottony-cushion scale was kept under control in its native Australia by some parasitic insect. He then reasoned that the scale outbreaks in New Zealand and California had occurred when the insect was imported without its parasite, allowing scale populations to grow unregulated by natural enemies.

Riley and the California Fruit Growers Association wanted to send a

USDA entomologist to Australia to collect potential parasites of the cottony-cushion scale but were caught in a governmental bureaucratic bind; the U.S. government would not fund foreign travel for USDA employees. Riley circumvented that regulation by sending one of his assistants, Albert Koebele, to the 1888 International Exposition in Melbourne, Australia, as a representative of the State Department. Koebele spent little time on State Department business, but did travel through much of Australia collecting a fly parasite that he found living inside, and feeding on, cottony-cushion scale insects. He also discovered a predacious lady bird beetle named *Vedalia* (now *Rodolia*) that fed on these scale insects, and had an exceedingly large appetite.

Koebele shipped the beetles and fly parasites he collected to California, and they were released in canvas tents surrounding orange trees infested with cottony-cushion scale. The beetles and the fly parasites quickly consumed all available scale insects, and when released from the cages they dispersed on their own and with human assistance all over California. Within weeks, the cottony-cushion scale insect was history, and from 1888 on its populations have been kept under control by the combination of fly parasite and beetle predator.

Riley cited the eloquent testimony of William Channing, an orange grower from Pasadena, California, in his 1893 Annual Report: "We owe to the Agricultural Department the rescue of our orange culture by the importation of the Australian lady-bird, *Vedalia cardinalis*. The white scales were encrusting our orange trees with a hideous leprosy. They spread with wonderful rapidity and would have made citrus growth on the whole North American continent impossible within a few years. It took the Vedalia only a few weeks absolutely to clean out the white scale. The deliverance was more like a miracle than anything I have ever seen." The entire project cost less than $2,000, not including the cost of a gold watch presented to Koebele and a pair of diamond earrings presented to Mrs. Koebele by grateful fruit growers.

The outstanding success of the cottony-cushion scale project stimulated an onslaught of foreign travel by pest managers searching for magic bullets to use against other imported pests. By 1900 there was a great increase in the search for control agents, particularly for use against insect pests that had been imported to North America, and especially to

California. Entomologists were traveling all over the world, to the native regions of imported pests, and searching for natural enemies. Back home, state and federal governments were constructing quarantine facilities to house and test the legions of proposed natural enemies that were flowing to North America. Propagation facilities also were built to raise copious numbers of control agents for distribution and release. The frenetic pace of biological control peaked between 1920–1940, when 85 major projects were undertaken to discover, test, and release biological control agents for use against North American insect and weed pests.

These world-girdling entomological explorers were looking for a variety of natural control agents, including predators, parasites, and diseases, and quickly focused on a few types of organisms that seemed to have the greatest potential for success. For example, the solitary parasitoid wasps have long egg-laying tubes called ovipositors protruding from their abdomens, with sensory hairs and cells at the tip that can smell and taste potential hosts. The female wasps insert their ovipositors into other insects and lay eggs, which develop into larvae that usually feed internally on the host. The host pest eventually dies, and the larvae complete their development into adult insects to continue the cycle. Each parasite species usually is limited to one or a very few host species, and their specificity makes them highly attractive as control agents.

Predator species popular among the entomological adventurers included ground and lady beetles, lacewings, robber flies, assassin bugs, and many other insects that feed on other insects and seemed to have good potential for biological control. The search was not limited to insects, however. Pest managers returned from their journeys with fish that fed on mosquito larvae and tiny predacious mites that fed on plant-feeding mite pests. In addition, a diversity of weed-eating insects, nematodes, and mites were discovered and imported for use in weed control programs.

Disease-causing pathogens were not neglected in the search for biological control agents. Bacteria, viruses, fungi, protozoans, and rickettsiae were isolated from pests in their home countries, imported to North America, cultured, tested, and formulated for use as microbial insecticides. Those with the greatest success included the *Bacillus* bacteria, similar to those sprayed against the gypsy moth in Vancouver, and the

NPV viruses, nuclear polyhedrosis viruses that turn insect larvae into dark bags of rotting protoplasm.

There is no doubt that some of these imported biological control agents were successful. Klamath weed is a good example of a pest whose virtual eradication can be attributed to biological methods. This European weed, also called St. Johnswort, earned its North American name because it was first reported near the Klamath River in northern California. By 1944 it occupied over two million hectares of rangeland throughout the United States and Canada, out-competing local forage plants used by free-ranging cattle and sheep and causing illness in livestock. The Australians had dealt with the problem by importing three European species of plant-feeding beetles that specialized on Klamath weed. American entomologists received authorization in 1944 to import these beneficial beetles from Australia, since World War II made it impossible to import them from their native Europe. Two of the beetle species became established in North America, and one in particular proved devastating to Klamath weeds. Within ten years, weed abundance was reduced by 99 percent, resulting in a continuing $3.5 million annual benefit from an original investment of less than $300,000 (values are in 1950s dollars, and so these benefits would be higher in today's dollars).

Or take the citrus blackfly in Mexico, Texas, and Florida. This insect is from Asia and first appeared in the New World in the Caribbean, where it decimated the Cuban citrus industry. These tiny insects feed on the sap of citrus trees, resulting in leaf atrophy, tree defoliation, and complete crop failure within a year of infestation. The Cubans had great success in controlling citrus blackfly with an imported Asian parasitoid wasp, but this wasp species was not effective in the southern United States and Mexico because of its inability to thrive in the high-humidity environments typical of citrus-growing regions on the North American continent.

Two species of parasitoid wasps from India and Pakistan were then found and imported to North America in the late 1940s. This project became massive. At one time, 1,600 people were employed to rear and release hundreds of millions of these parasites in Mexico every year, thereby slowing the blackfly's spread into the United States. Subsequent infestations in Florida and Texas did develop but were controlled by similar parasite releases, preventing millions of dollars in annual crop losses.

Textbooks are filled with many other examples, but all are concentrated in the early to middle parts of the twentieth century. The reality of our experience with natural control agents is that despite some brilliant success stories, most biological control attempts have failed, especially in recent decades. Even with the failures, biological control has proven to be an important part of modern pest management, but the failures provide interesting instruction as to the limits of our ability to use nature to manage nature.

Opinions about failure rates and cost-benefit analyses of natural enemies depend on what data set you examine and what level of control is considered to be "successful." The overall world-wide rate at which imported biological control agents became established in their new environments is 30 percent of attempted introductions, with 103 of 267 weed control projects (38 percent) and 1,251 of 4,226 insect pest control projects (29 percent) resulting in the imported control agents surviving in their new habitat.

But all biological control agents that became established were not necessarily effective in controlling the pest for which their introduction was intended. Data on success rates for biological control agents controlling both insect and weed pests hover at around 16–18 percent of attempted introductions world-wide. These data are inflated, however, since many of these success stories involve repeat introductions of the same successful natural control agent. Also, "success" is defined broadly to include not just projects in which a pest was eradicated or reduced to levels below economic significance but also those which merely managed to reduce the necessary number of pesticide applications.

Cost–benefit analyses are more positive than the success rate data, since the relatively few successful projects yielded significant gains in crop production. One estimate of the dollar value for crops saved by biological control agents in California from 1928 to 1979 came in at an accumulated total of $987 million dollars, with costs of about $1 million for each success. Another estimate for California from 1888 to 1985 was $500 million in accumulated savings, less than the first estimate but still substantial. If we look only at success stories, the dollar ratio of the benefits of successful biological control measures compared with the costs is about 30 to 1.

But another set of statistics is perhaps more telling than data on success rates. Between 1970 and 1981, only 14 biological control experiments were attempted, as compared with 57 attempts between 1930 and 1940. Of those experiments attempted since 1970, only 4 were outstanding successes, the last one being in 1981. Overall, the effectiveness of biological control agents decreased from 36 percent for releases made between 1940 and 1949 to only 14 percent for those made between 1970 and 1979. Clearly, something has happened in recent decades that has put a damper on both attempts and successes in the biological control field.

The major "something" that happened was pesticides, which can harm natural enemies in addition to the pests themselves. This dilemma in pesticide use was recognized early on by Riley. In 1893, toward the end of his career, he wrote: "Occasions often arise when it were wiser to refrain from the use of insecticides and to leave the field to the parasitic and predaceous forms . . . The insecticide method involves the destruction of the parasitic and predaceous species, and does more harm than good." Rachel Carson also recognized this problem 70 years later in *Silent Spring:* "Sometimes the result of chemical spraying has been a tremendous upsurge of the very insect the spraying was intended to control . . . At other times spraying has let loose a whole Pandora's box of destructive pests that had never previously been abundant enough to cause trouble."

The destruction of natural enemies, whether native or imported, has induced numerous pest outbreaks and prevented biological control from becoming a significant factor in managing field pests. In cotton, for example, pests such as bollworms, budworms, aphids, spider mites, and loopers all have reached outbreak proportions due to pesticide-induced destruction of their natural enemies. In apples, pest outbreaks following spraying have included the European red mite, red-banded leafroller, San Jose scale, oystershell scale, rosy apple aphid, wooly apple aphid, white apple aphid, two-spotted spider mite, and apple rust mite. Similar lists of pests that have erupted following spraying because their natural enemies were reduced can be found for most of our agricultural crops.

Ironically, the earliest recognition that pesticides could disrupt pest control by natural enemies came from California, about the same time that the cottony-cushion scale control program was being heralded as a stunning success. A predacious beetle similar to *Vedalia* was introduced in 1892 to control black scale on olive trees. In 1893 kerosene emulsion was used in Santa Barbara against black scale. Farmers noticed that the predacious beetles were nowhere to be found in sprayed areas but were abundant and effective away from the sprays. In contrast, the black scale was abundant in sprayed areas but less abundant in the unsprayed olive orchards where the beetles could be found. The conclusion, of course, was that spraying had reduced the beetle populations, allowing the black scale to increase in sprayed orchards.

Even the most classic and successful case of biological control, the cottony-cushion scale, has not been immune from pesticide effects on natural enemies. In the 1940s citrus growers began experimenting with DDT and other highly toxic new pesticides. Suddenly, populations of the *Vedalia* beetles and fly parasites were virtually destroyed in many parts of California, and the scale pest began to reappear. The natural control agents had to be reintroduced to many regions, and only regular and repeated releases coupled with minimal spraying allowed these predators and parasites to reestablish control over the cottony-cushion scale.

The Cornell University entomologist David Pimentel and his colleagues calculated that $520 million in annual crop losses are caused by pesticide reduction of natural enemies in the United States. In some situations, pesticide effects on biological control agents can be catastrophic. For example, insecticide use in Indonesian rice production in the early 1980s destroyed natural enemies of the brown planthopper, and the populations of this pest then exploded. The Food and Agriculture Organization (FAO) estimated that $1.5 billion in rice production was lost in just two years. Fortunately, Indonesian President Suharto followed the advice of his specialists and ordered severe reductions in pesticide use, which allowed the natural enemies to increase and brought the pest levels back down below economically tolerable thresholds.

Perhaps the best assessment of the failure of managed biological control to compete with pesticides has been made by the marketplace. There is no private industry in North America, or anywhere in the world, that supplies any significant quantity of natural enemies for control of field crops. Biological control remains in the domain of the research scientist and government employee, with use by organic farmers and greenhouse growers providing a very limited market for commercially produced biological control agents.

Ninety-five small companies in North America produce and sell beneficials. These companies have natural, down-homey, organic-sounding names like Peaceful Valley Farm, Organic Pest Control, Natural Pest Controls, Better Yield Insects, and Beneficial Biosystems. They are peewees in the business world of pest management, however. Maclay Burt, President of the Association of Natural Bio-control Producers, estimates that annual world-wide sales of beneficial insects are "$50 million a year, and that's stretching it." A sales volume in the $1 million a year range would elevate a company into the upper stratosphere of the biological control industry.

Compare that with the 1,200 U.S. companies producing $8.5 billion in pesticides each year. These chemical companies are dominated by a few major players that each sell product valued in the hundreds of millions of dollars annually, with hard-sounding corporate names like Dow-Elanco, American Cyanamid, and Technicide. In name, size, and impact, commercial biological control provides only a thin alternative to our pesticide-dominated industries.

Indeed, the only economically significant markets for sales of beneficial control agents are indoors, in greenhouse and interior plantscape management, and outdoors on a few specialty crops such as strawberries and oranges. Biological control companies all carry similar product lines consisting mostly of beneficial insects and mites for control of greenhouse pests such as whiteflies, spider mites, flower thrips, aphids, mealybugs, and scale insects. Most of these companies sell the same species of predators and parasites, and the introduction of a new biological control agent is an unusual occurrence that can provide a significant marketing edge. There are only about 20 commonly used commercial biological control agents sold world-wide. In contrast, as of 1993 there were 860 chemical pesticide products registered for use in the United States, and 20 or so are added each year.

❧

Biological control definitely is not big business. It is, however, an interesting business, and one that requires considerable sophistication in production facilities and marketing. One of the most successful of these biological control businesses in North America is Applied Bio-nomics, located on the scenic Saanich Peninsula of Vancouver Island, British Columbia. This is the warmest part of Canada, with a Pacific Northwest coastal climate that rarely sees snow. The Saanich area is a retirement haven for well-off Canadians fleeing the bitterly cold Canadian prairie winters, and it also attracts hobby farmers growing exotic crops such as kiwi fruit, unusual varieties of apples, and every type of organic produce one can imagine. Farm markets, with labels like Island Fresh, are everywhere on the island.

Applied Bio-nomics fits perfectly into this high-end lifestyle. Its office is a renovated 70-year-old farmhouse nicknamed the Bug House, located on five acres off a rustic country lane, surrounded by rolling hills overlooking the island-dotted Straits of Georgia. Posters proclaiming "Entomologists Get the Bugs Out" hang in almost every room, and books like *Common Sense Pest Control* and *Personal Organization* sit in government-issue wooden cabinets. The country setting looks like the West Coast equivalent of a picture from an L.L. Bean catalogue, but Applied Bio-nomics is much more than its superficial ambience. Rather, this is a highly refined business, with computer-controlled climates in 13 production greenhouses and a staff of 12 with considerable expertise in developing, producing, and marketing biological control products.

The guru of Applied Bio-nomics is a former high school teacher named Don Elliot, whose early love for insects was generated in the steamy Amazon jungles of Colombia where his father worked for an oil company during Don's youth. Elliot is fiftyish and rumpled, and affects a professorial air of distracted befuddlement. However, Elliot is anything but befuddled. He runs a very sophisticated business, and beneath his frazzled appearance lie the heart and skills of a Wall Street entrepreneur.

Applied Bio-nomics began in the late 1970s, when Elliot and his wife moved to Vancouver Island looking for a change of habitat. His interest in bugs led him to develop a joint project with Agriculture Canada and the British Columbia provincial government to produce predators and parasites that could control pests in greenhouses. This work was stimulated by serious outbreaks of pests on crops grown under glass that required increasingly frequent and toxic pesticide sprays. The growers were worried, because the pests quickly became resistant to the pesticides, and the quantities of chemical used left residues that made the crop unmarketable.

Elliot's first project involved a predatory mite called *persimilis*, a voracious consumer of another mite, the spider mite, that is an equally voracious consumer of many high-value greenhouse crops such as tomatoes, cucumbers, and peppers. The persimilis mite is one of the most common biological control agents produced for sale in North America, but the rearing, propagation, and successful use of this beneficial insect are anything but simple.

Applied Bio-nomics uses three separate greenhouses to produce the predatory persimilis mite. First, bean plants are grown in one greenhouse that must be kept free of all pests to allow good plant production. Then, the beans are moved to a second greenhouse where the pest spider mites are maintained as a food source for the persimilis. When the beans are well-infested with spider mites, the plants are moved to yet a third greenhouse where the persimilis are introduced to feed on the spider mites and then reproduce. Finally, the persimilis need to be scraped from the bean leaves, separated from any remaining spider mites, and quickly shipped around the world for release in spider mite-infested greenhouses. None of this is cheap; a single greenhouse costs about $400,000, and quality control must be rigorous, as contamination at any stage will set the production line back by weeks or months.

Further, production methods for known agents as well as the development of new predators and parasites must be kept highly confidential. As Elliot put it, "The only way to protect technology is to protect development. You can't patent any of these systems, and there's a huge jump from laboratory production to the field, generally taking five or more years." Success can be profitable, or at least as profitable as a small-scale business with a limited market can be. Where field crop farmers consider $50 per acre to be a considerable expense, the high-value greenhouse industry will spend $5,000 for each acre under glass and not consider that unreasonable.

Applied Bio-nomics has been successful at developing and marketing both known and new products, and today it is one of the world's largest biocontrol companies. Its product line is unusually large, and includes about 20 different predators and parasites with names like the Aphid Preventor, Greenhouse Spider Mite Eliminator, and Spider Mite Predator. Elliot's client base is diverse and cosmopolitan, including greenhouses, shopping malls, zoos, indoor gardens, and Epcot Center's Agriculture Pavilion. Even so, annual sales are just below $1 million, and the corporate trappings of Applied Bio-nomics are definitely small-business.

I asked Elliot what concerns he had about the future of his business, and the entire biological control industry. I expected a diatribe against pesticides, a detailed exposition of how difficult it can be to learn how to rear a new biological control agent, or perhaps concern about something as simple as a power outage killing his products by upsetting the rigorous greenhouse climates he must maintain. Surprisingly, he complained about the same problems that the pesticide empire routinely complains about—too much government regulation and environmentalists hamstringing his business:

> Anything that eats something else, from a grower's view they want it. It's worrying us that Agriculture Canada and the U.S. Department of Agriculture are jumping into the industry before it's developed and trying to regulate it. It's all up in the air right now. They tried to get some laws in place and it was a disaster. They would have closed the entire industry down. They did stupid things like require us to put "keep away from children" on each bottle. Their whole mode is that they have to regulate something, and the only model they have is pesticides. The existing regulatory system can't control the use of illegal, non-registered products, and they're going to have a pretty hard time with biocontrol agents . . . Believe it or not, Greenpeace and other environmental groups complained because our products were too broad-spectrum. Let's get sensible; you've got to look at the risk–benefit of this.

Two groups, the International Organization for Biological Control and the Association of Natural Biocontrol Producers, are attempting to represent this industry in the regulatory corridors of government. It is ironic, however, that our society has mandated pest managers to find alternatives to pesticides, but we are regulating natural alternatives to the extent that lobby groups from industry have formed to combat those regulations.

Indeed, ironies abound in the beneficial insect field. It seems to be a particularly cruel twist of fate that our impact on our environment has been so extensive that we must substitute commercially grown organisms for natural predators and parasites, and for feral pollinating bees. We made the decision to clear-cut the beneficial insect world and then erected in its place expensive managed systems, often with deleterious side effects.

While we marvel that our human ingenuity can invent managed pollination and release cultured predators and parasites into greenhouses, our own pest management follies forced us to create these managed substitutes in the first place. Somehow the sound of a spray plane dumping chemicals on our crops doesn't quite substitute for the natural buzz of feral bees, the elegant hum of a parasitic insect searching for a host, and the sound of predators crunching pest insects and mites in their jaws. We have fallen into an agricultural system that is dependent on increasing levels of management to solve problems that we ourselves created. We have lost something real here, something that we don't recognize when we bite into an apple, eat a fresh loaf of bread, or put on an all-cotton shirt.

What we have lost is nature, although we have substituted advertising images of wholesome-looking farm scenes to delude ourselves into believing that our food and clothing products still come from a pristine natural world. Not so; the reality of agriculture is heavily dependent on pesticides and management. The "natural" way of doing agriculture that beneficial predators, parasites, and pollinators represent is a relic. Whether it has to be this way is an open question, but beneficials do remind us that there may be alternatives to the managed systems we have constructed to replace nature.

CHAPTER EIGHT

Frankenstein Plants

"In the long term (10–50 years), unforeseen ecological consequences of using recombinant organisms in agriculture are not only likely, they are probably inevitable. But it is crucial to put this into perspective: it is difficult to describe a credible scenario that will lead to a problem that is different in kind from the problems caused by, and grappled with, in past agricultural practices."

Office of Technology Assessment, U.S. Congress (1988)

There is perhaps no better measure of the rate that science has progressed in the twentieth century than the fact that we did not even know the structure of DNA until 1952. Today, only 45 years later, we not only are well on our way to mapping every one of the multibillion components that make up human DNA, but we are taking genetic material from one species and inserting it into another. This novel technology has inspired extreme reactions and hyperbole on both sides of the fence—from those who believe that our ability to manipulate genes will lead to a world free of pests and diseases, to those convinced that we are creating health and environmental threats of unprecedented proportions.

The field of pest management has not been exempt from these debates about the potential, for good and ill, of this new technology. At the cutting edge of pest control today are plants whose genomes have been bioengineered with inserted genes from bacteria, viruses, and other plants. These foreign genes instruct their new host plant to produce proteins that kill insect, fungal, and viral pests and diseases, or prevent crops from being damaged by harsh chemical herbicides put down to control competing weeds. While pheromones and other biological methods of pest control languish in the relative commercial obscurity of alternative technologies, these molecular techniques have led to some of the most economically significant new products in pest management since the advent of synthetic chemical pesticides.

The rapid growth of this new industry has not been without its detractors. The terms "transgenic," "recombinant," and "bioengineered," which describe the organisms created by molecular biology, conjure up our worst fears about the dangers of technology run amuck—fruits, nuts, and vegetables that may poison or induce disease, unnatural Frankenstein plants that may invade wild plant populations and create new varieties with unimaginably horrible environmental consequences.

Ignoring these doomsday concerns, the corporate and agricultural communities have invested enormous sums of research time and development money into recombinant plants for the purpose of controlling pests and diseases in plants. Companies like Monsanto, Dow-Elanco, Rhone-Poulenc, Sandoz, Ciba-Geigy, Rohm and Haas, Dekalb, Pioneer Seeds, and many others are competing for this new market. Transgenic crops such as corn, cotton, potatoes, soybeans, tobacco, and tomatoes have reached the commercial stage, and extensive acreages all across North America have been planted with these recombined organisms.

Pest managers are gambling that they can make this new technology work without serious health and environmental side effects, while opponents of transgenic plants are convinced that we are propagating the worst pest management disaster since DDT. In the middle are the regulators, government agencies responsible for ensuring that human and environmental health are not threatened by new technology. After all, if there is one thing that chemical pesticides should have taught us, it is that miracle cures often come with a price.

What is most interesting about the transgenic plant debates is that bioengineered plants can substantially reduce the use of that already known hazard, chemical pesticides. The main challenge for government regulators is to interpret correctly the possibility of deleterious impact from bioengineered products and to weigh that risk against the beneficial effects of reducing pesticide use. Our pesticide experiences have taught us that government regulatory agencies can sacrifice public health and environmental safety for industrial and commercial interests. The key question we need to ask, and answer correctly, is whether government regulation of bioengineered plants has struck the proper balance between safety and efficacy. Are the protesters right, is industry right, or is there a middle ground in which society's best interests lie?

The genetic techniques behind this debate are conceptually simple, although in practice not as easy to implement as molecular biologists would lead us to believe. Practitioners of the recombinant craft remind us that there is nothing new about manipulating DNA. Nature has been recombining genes since life began, and plant and animal breeders routinely interfere with the "natural" course of evolution by selecting individual plants and animals that have various beneficial characteristics and breeding them until a desired variety is achieved.

This oversimplification perhaps provides some comfort to a nervous public, but it masks the drama actually unfolding in our molecular laboratories. For the first time in the history of life, we are able to selectively choose individual genes from one organism and put those genes into another vastly different kind of organism. Whereas breeders, and nature itself, must work with closely related organisms that can be crossed or hydridized, bioengineers are not restricted in this way. They can, for example, put human genes into organisms ranging from pigs to bacteria. There are few technical limits to the methodology used to perform these gene transfers, although in practice it may take five to ten years to bioengineer the desired product. Only ethics and a sense that at least some extreme recombinations may be immoral or dangerous limit how far we will go in creating new transgenic organisms.

The basic techniques used to create transgenic organisms can be found in any high school biology textbook, although the terminology in most descriptions for the lay public tends heavily toward analogy. It relates how molecular "scissors" snip genes from donor organisms at specific sites on the DNA molecule, and then other molecular tools "sew" or "paste" the genes into the proposed host's genome. The actual techniques are logistically more difficult that these metaphors would suggest, but molecular biologists essentially do cut pieces of DNA from one organism and incorporate them into the genetic code of a different species. These recombinant genes are then replicated every time a cell divides and can be passed on indefinitely from one host generation of cells to the next.

Plants will accept transferred genetic material invoking resistance to pests, diseases, and herbicides. Many plants already have a considerable natural arsenal of defenses against attack, including an impressive array of chemical defenses such as THC (the active ingredient in marijuana), cocaine, nicotine, strychnine, cyanide, and many others. Nicotine, in fact, was one of the first chemical insecticides used in modern agriculture. These defensive substances evolved over millions of years in order to deter insects from feeding. Some insects have succeeded in countering the chemical defenses of plants with their own chemical mechanisms to detoxify plant-produced compounds, but this process of coevolution has been slow enough that the corresponding plants can survive, often evolving yet newer mechanisms to counter successful insect feeders.

The beginnings of molecular biology coincided with an increased understanding of the chemical nature of plant defenses. Thus, it was not surprising that the earliest biotechnologists recognized the potential to introduce biochemical defenses into plants by transferring genes from one organism to another. This work has proceeded in both industry and academia, despite controversy. Although some of the earlier public protest that accompanied this and other recombinant DNA technologies has subsided, there is still a substantial protest movement against bioengineered organisms. However, a decade of experience in highly controlled experiments has convinced regulatory bodies that the risks involved in propagating transgenic plants are small, and the potential benefits substantial.

Some of the most successful transgenic products to date involve genes from *Bacillus thuringiensis,* the bacterium used to control gypsy moths in Vancouver and elsewhere. *B.t.* is a common soil-dwelling organism whose toxins are inconsequential for vertebrates but, in some insects, disrupt gut membranes under the highly alkaline conditions found in their digestive system. These toxins evolved to allow bacteria to feed on the nutrients released by the broken gut cells. After the insect dies, within two to three days of infection, the bacteria return to the soil to await another hungry host. Different strains of *B.t.* produce proteins that are active only against specific groups of insects.

Bacillus thuringiensis products are the most frequently used "natural" biological control on the market today, but current formulations have to be sprayed on crops or forests. Thus, a *B.t.* spray is active for only a short time due to its instability under field spraying and exposure conditions and therefore is not always effective at getting into the insect gut. Also, sprayed formulations of *B.t.* are expensive.

A better way to deliver *Bacillus thuringiensis* toxins to pest insects, according to biotechnologists, is to insert the bacterial genes producing these toxins directly into the plants on which the insects feed. A number of these new toxin-producing crop varieties have entered, or are poised to enter, the marketplace, including corn, cotton, potatoes, and tomatoes, with others close to becoming commercial. Each transgenic crop contains genes producing the *B.t.* toxin that is most effective against an important pest of that crop, but otherwise the crop is identical to non-transgenic varieties of the same plant. These new products have optimistic names such as Yield-Gard Insect-Protected Corn, Mycogen Corn with NatureGard Gene, or The Maximizer: Hybrid Corn with the Knockout Gene. Whatever the names, pest insects feeding on these transgenic plants are quickly poisoned, to the delight of farmers who no longer have to spray their fields with insecticides.

Another type of transgenic plant reaching the marketplace today contains genes conferring herbicide tolerance. Most herbicides act by inactivating proteins that are found only in plants, or in more limited plant groups, or even in particular species. Many herbicides are broad-spectrum, however, and harm crops as much as weeds, since most crop plants contain the same protein systems as do their pest weeds. Thus, a bioengineered crop that would resist herbicides would allow weed control to proceed without harming a crop that was in the field already. Although success in providing herbicide resistance to a crop would increase herbicide use, this does not appear to be a major concern to farmers interested primarily in crop production.

Three basic approaches are being used to produce transgenic, herbicide-tolerant crops. Since herbicides act by binding to and inactivating specific plant proteins, one approach has been to bioengineer plants to produce proteins that resist herbicide binding but still perform their vital functions for the plant. A second approach is to insert genes into plants that overproduce the target protein, so that there is surplus available even after herbicide use. Finally, some transgenic plants have been produced that contain new enzyme systems to degrade or detoxify herbicides.

The first bioengineered herbicide-resistant plants were tobacco and oilseed rape, in which mutants of two bacterial species were used as the source of resistance-inducing genes. The target herbicide in these projects was glyphosate, commonly known as Roundup. This herbicide interferes with amino acid synthesis by blocking the action of a particular enzyme. The mutant bacterial genes either alter the enzyme so that glyphosate cannot bind to it, or else overproduce it so that there is a surplus in the crop plants even after herbicide use.

An enormous amount of research and development money has gone into producing these new transgenic varieties, but the economic benefits are high enough to justify the expense. By 1993, 61 public institutions such as universities and government laboratories and 88 private companies were involved in agricultural biotechnological research. Field trials of transgenic crops began in 1986, when the U.S. government instituted a well-defined system to review and approve trials outside the laboratory with transgenic plants. At least 2,750 field trials were conducted world-wide with transgenic plants between 1986 and 1995, with about 70 percent of those trials in the United States and Canada. Potato has been the most heavily tested crop, with oilseed rape, tobacco, corn, and tomato rounding out the top five crops being tested. Others being examined in field trials include soybeans, cotton, flax, squash, and rice. In 1993 the procedures to approve field trials were simplified, and in 1994 full-scale commercial trials for pesticidal varieties of corn, potato, and cotton were approved by the Environmental Protection Agency.

This research has been expensive; industry alone spent about $250 million dollars on biotechnology research in 1993, with one company, Pioneer Hi-Bred, investing $100 million. Industry funding has been supplemented by federal funds; in 1994, for example, federal government agencies spent $234 million on biotechnology research. The entire federal budget for research into all other alternative pest management techniques that year was a mere $165 million.

The economic potential for agri-biotechnology was enhanced during the 1980s by rulings of both the U.S. Supreme Court and the Patent and Trademark Office that allowed transgenic plants and seeds to be patented. The market for pest- and disease-resistant plants is expected to reach a value of $12 billion by the year 2000, with herbicide-resistant crops following close behind at $6 billion. At this rate, the commercial value of transgenic crops will soon surpass that of pesticide sales.

A close look at one transgenic crop, NewLeaf Russet Burbank Potatoes, indicates just how attractive transgenic plants can be to farmers, relative to pesticide products. The NewLeaf potato incorporates *B.t.* genes that produce compounds toxic to the Colorado potato beetle, the most devastating insect to the North American potato crop. According to product information put out by NatureMark, the division of Monsanto that markets NewLeaf potatoes, this transgenic variety provides 100 percent control of beetles, with no need to use any chemical insecticide. The protection is evenly spread throughout potato fields and is effective for the entire season, since it is incorporated within each plant. In contrast, pesticide sprays are effective for only a few days after spraying and provide spotty control owing to wind, weather, and spray equipment malfunctions. Even better, the NewLeaf potato has no effects on any other organisms, including predators and parasites of the Colorado potato beetle and other potato pests. Finally, there are no concerns about worker safety or environmental impact from NewLeaf potatoes, at least as far as NatureMark is concerned.

The bottom line, however, is cost, and here the figures provided by NatureMark are impressive. The cost of controlling the Colorado potato beetle with chemical insecticides can go up to $300 acre, and these insects have developed resistance to most of the pesticides licensed for use against them. The extra cost to purchase NewLeaf Russet Burbank seed potatoes containing the antibeetle toxins is only $22.50 per acre more than conventional Russet Burbanks. Further, yields increase with the NewLeaf variety, since damage from the beetles is eliminated.

Similarly favorable cost–benefit analyses are available for other transgenic products, such as *B.t.* corn varieties, which are protected from corn borer, the major insect pest of corn in North America. Savings from the use of *B.t.* corn are predicted to include $50 million in reduced insecticide costs and $1–1.5 billion in increased yields, based on reductions in current 5–7.5 percent yield losses attributed to corn borer.

❧

Industry's enthusiasm for transgenic products has not been shared by a small but active cadre of environmentalists, geneticists, and ecologists who have registered concerns about these products from the onset of biotechnology in the mid-1970s. Three groups have emerged as the major organized critics of transgenic plant biotechnology: the Union of Concerned Scientists, the Council for Responsible Genetics (both based in Boston), and the Foundation on Economic Trends (in Washington, D.C.). Each of these groups has done extensive research into the potential risks of transgenic plants and published numerous books, reports, newsletters, and articles expressing their concerns. In addition, they have initiated lawsuits designed to prevent field tests of transgenic plants, which have delayed but ultimately not prevented the tests from proceeding.

The "mediators" in this battle have been the traditional bodies that advise government concerning science policy, including prestigious U.S. science organizations such as the National Research Council, the National Academy of Sciences, and the National Agricultural Biotechnology Council. In addition, the U.S. government has issued reports through the Office of Technology Assessment, the Environmental Protection Agency (EPA), the Food and Drug Administration (FDA), and the Department of Agriculture (USDA). The major U.S. target of this intense activity has been Congress, which legislates if and how transgenic plants can be used. Ultimately, of course, public opinion has been the focus of lobbying efforts, because public pressure eventually determines the directions that biotechnology regulation will take.

Concerns about transgenic plants focus on four major areas: (1) human health risks associated with food contamination, (2) the potential for the transplanted genes to jump from crop plants into weeds, (3) increased herbicide use that would result from planting herbicide-resistant crops, and (4) pest resistance to transgenic organisms. The exchanges between environmentalists and industry have taken on ritualistic overtones, with one side routinely expressing concerns about the health and environmental risks of recombined organisms, and the other just as routinely denying the problems and supporting the economic benefits of transgenic plants. Government regulators end up in the middle, having to make the decisions concerning whether a field test or product sale will proceed.

Risks to human health through food contamination by bioengineered products were the most significant concerns at the start of the transgenic debates in the late 1970s. Even today, opponents of food containing recombinant genes tend to the extreme, as this excerpt from *Health Action* (October 1993) illustrates:

> In the movie *Jurassic Park,* scientists use dinosaur DNA recovered from a drop of ancient blood and create a herd of prehistoric monsters. The story is fictitious, yet the facts surrounding today's biotechnology industry are more frightening than the scariest scenes that Stephen Spielberg could ever imagine. Scientists are fooling with natural evolution and tinkering with genes between organisms and species to incorporate particularly desired traits. These transgenic products, as they are called, now include mixing tomatoes with fish genes, potatoes with moth genes, and pigs, cattle, sheep and fish with human genes. These mutant Frankenstein foods will be coming to a supermarket near you and if the industry has its way, you won't even know it.

In spite of this inflamed opposition, regulatory agencies are not overly concerned about health risks associated with transgenic plants. Experience with biotechnology and extensive testing of transgenic plants has given them a clean bill of health. The most general concern, release of new infectious microbes, has proven unfounded throughout the entire biotechnology field, and even the most rabid opponents of recombinant DNA no longer concentrate on this issue. In addition, no lethal or even sublethal impact of transgenic plants eaten by nontarget organisms has been found, suggesting they are safe to consume. All potentially commercial transgenic plants are exhaustively tested using dietary and toxicity studies on nontarget organisms, including mice, rats, and birds as well as beneficial insects such as honey bees, predatory beetles and lacewings, and parasitic wasps.

The lack of toxic effects is not surprising, since there is nothing about the gene products produced by transgenic plants that would suggest toxicity to people or animals. The novel proteins produced by transgenic plants are treated just like any consumed protein and are rapidly inactivated and degraded in stomachs by digestive enzymes. In addition, nutritional studies have not found any differences in the levels of protein, fat, carbohydrates, or fiber content between transgenic and conventional plant varieties. Thus, no general labeling requirements have been imposed on transgenic plants in supermarkets, since they are considered identical to conventional varieties as food products.

The one exception, and an area of more substantive concern, is the potential for the transfer of allergenic substances that could induce fatal allergic reactions in susceptible individuals. The likelihood of allergic reactions to the proteins transferred into plants for pest management purposes is low, but a recent study by Julie Nordlee and colleagues from the Food Sciences and Technology Department of the University of Nebraska indicated a potentially serious problem with transgenic soybeans.

Soybeans are deficient in the amino acid methionine. Scientists working for Pioneer Hi-Bred International developed a variety of transgenic soybeans that is rich in methionine by taking genes from methionine-rich Brazil nuts and incorporating them into soybeans. However, the Nordlee study demonstrated that this bioengineering also transferred the Brazil nut proteins that induce allergic reactions in susceptible people, so that someone allergic to Brazil nuts could have had a serious or even fatal reaction after consuming the transgenic soybeans.

According to current FDA policy, Pioneer Hi-Bred soybeans or their products would be required to carry a label indicating their potential danger to allergic consumers. Instead, the company decided to discontinue plans to market this product. While this case indicates the need for caution with food products from transgenic plants, it also demonstrates that current regulatory protocols can be successful at preventing allergenic food products from reaching the marketplace unlabeled.

A second potential problem with transgenic plants is the possibility that negative ecological consequences would result from some traits being incorporated into crops, particularly herbicide tolerance. The concern here is that an herbicide-tolerant domesticated crop plant could revert to a wild, weedy state and become a pest. In addition, environmentalists fear that tolerant genes could "jump" from a crop into a closely related wild plant, creating superweeds. Either situation could pose serious consequences for both agriculture and the environment, and opponents of transgenic plants such as Jane Rissler and Margaret Mellon of the Union of Concerned Scientists believe that we cannot yet predict the consequences of using these crops in the field: "The long-term impacts of transgenic plants in the environment are unpredictable and difficult, if not impossible, to predict at the current level of scientific understanding . . . No company should be permitted to commercialize a transgenic crop in this century until a strong government program assuring risk assessment of all transgenic crops is in place."

Again, however, regulatory authorities have paid considerable attention to this potential problem, and the risk assessment demanded by Rissler and Mellon has been done. Regulators have come to believe that herbicide tolerance will not be a problem if we carefully select which crops to bioengineer. The issue of whether a crop plant can go native and become a weed is simply not a concern among pest management professionals. Crops are so heavily selected for cultivation that they retain few weedy characteristics such as thorns, seed dormancy, extensive root systems, and bitterness that are responsible for the success of feral weedy plants. Conventional breeding has modified crops to a much greater extent than has the addition of a single herbicide-tolerant trait, and yet these crops have not escaped cultivation and become pests.

The possibility that hybridization between genetically modified crops and wild relatives might enhance weediness could be a more serious objection to certain recombined plant varieties. Closely related plants often cross with each other by pollen from one species fertilizing the ovum of another, creating a route for transfer of a transgenic characteristic from a crop to a related feral weed. American and Canadian agriculture is somewhat protected from this potential danger, because most North American crops have been imported from elsewhere and have no close relatives with which to hybridize. A 1989 National Research Council report pointed out that "the paucity of crops derived from North American sources means there will be relatively few opportunities for hybridization between crops and wild relatives in the United States, except where both crops and wild relatives have immigrated. The incidence of hybridization between genetically modified crops and wild relatives can be expected to be low."

There is another, more compelling reason besides gene jumping to question the introduction of herbicide tolerance into crop plants, and here the objections of concerned scientists resonate more forcefully than for other concerns. The purpose of bioengineered herbicide-tolerant crops is to permit herbicide use while the crop is in the field, which would increase the use of already overused chemicals. This is particularly ironic, as pointed out by the Council for Responsible Genetics: "One of the early promises of biotechnology was that it would provide environmentally preferable alternatives to current agricultural practices, including alternatives to chemical pesticides. Instead, herbicide-tolerant plants promise prolonged chemical dependency."

Proponents of herbicide-resistant crops argue that these recombinant plants would bring both ecological and economic benefits by increasing the use of safer herbicides, allowing more precise applications, and reducing costs to farmers, but these claims ring hollow. For one thing, many crops being tested have been designed to tolerate some of the harsher chemical herbicides, such as bromoxynil, atrazine, and 2,4-D. It also is difficult to understand how costs associated with increased herbicide spraying, in addition to the higher price of transgenic seed, would save farmers money. Finally, it is not a coincidence that companies marketing herbicides, such as Monsanto, are among the major developers of herbicide-tolerant crops.

The objective of herbicide-resistant plants is not better pest management but higher pesticide sales. The trade publication *Agricultural Genetics Report* pointed out as early as 1983 that "many major firms involved in agricultural chemicals are developing crops resistant to their new experimental herbicides in the hope of selling the seed and chemicals as a pair. Others see herbicide resistance as a way of regaining market share lost after a well-known herbicide has declined in price and popularity. Again, the old herbicide is sold in combination with a new seed resistant to it." Zamir Punja, a Canadian expert on plant biotechnology who has worked in both academia and industry, described to me in 1996 the rationale behind the development of Monsanto's glyphosate-resistant transgenic soybeans: "It will increase use of the chemical, without a doubt, and yes, it's there to increase profits for the company."

A final controversy concerning transgenic plants engineered against pest insects is that these crops will increase the survival and reproductive success of insects that can overcome bioengineered plant defenses. Many transgenic crops have been designed to be an alternative to insecticides, but they may bring about one of the detrimental effects of chemical pesticides, which is the rapid evolution of resistance in target pests. The development of insect resistance could in turn lead to increased use of chemical pesticides, which would be detrimental to the environment. In addition, resistance would have significant economic impact on the ability of transgenic plants to replace pesticides.

The dilemma facing farmers in using transgenic crops is that transgenic organisms confront pests with classic challenges that are reruns of the pesticide treadmill story. If pesticides have taught us anything, it is that a few pest individuals will survive and their offspring subsequently will thrive following the application of strongly selective materials such as chemical toxins. The same evolutionary principles responsible for pesticide-induced resistance will act to make transgenic crops even more effective at inducing resistance.

Paradoxically, the high success rate of transgenic crops in killing pests is their most problematic trait. These crops are highly toxic to pests, which rapidly selects for those few individuals that can overcome transgenic defense mechanisms. Successful pests pass on their resistant traits to their offspring, and survivors of the initial transgenic attacks will have a field day feeding on the now-defenseless crops. Further, the selective power of transgenic plants will be even greater than for chemical pesticides, since there is no spatial or temporal escape for pests from these plants. Pests have no physical refuge from a field of transgenic crops, since entire regions will consist of transgenic plants expressing the same

toxic traits. In addition, the plants don't go away after a few days, as pesticides do, so that the selective regime will be relentless.

The transgenic industry is aware of the likelihood that resistance to transgenic crops will develop but believes that proper management of transgenic plantings can prevent resistance. Unfortunately, the specifics of "proper management" remain vague and may be difficult to implement in agricultural practice. I attended a symposium at the 1995 Entomological Society of America meetings that addressed this resistance problem and heard speaker after speaker deliver the industry line on this subject. It was clear that the producers of transgenic plants are greatly concerned about resistance, but the solutions were expressed in platitudes rather than concrete suggestions for resistance management. For example, one industry speaker typical of the group used phrases such as "product stewardship and education," "regular and consistent communication," "establishment of a mutually supported strategy," and "work together to solve issues." Nowhere in his talk, however, did I hear a tangible solution to the problem.

Fred Gould, an evolutionary biologist from North Carolina State University, described the essence of the transgenic resistance problem in a 1991 *American Scientist* article:

> Even if teams of academic scientists find solutions to the problems of pest adaptation, implementation of these solutions may be difficult. In the economically competitive environment of agriculture, the general orientation is toward short-term profits, whereas the concept of resistance management emphasizes long-term paybacks. In some cases, according to this approach, less is more. The lower the overall selective pressure challenging a pest, the longer the time for the pest to adapt. As a result of this approach, some pests may be left in the field. Even though these low densities of pests do not generally cause significant yield reduction, farmers may not be willing to take even small short-term risks for the promise of long-term stability of yield.

The solution to resistance may lie in using bioengineered plants that do not overkill pests. In practical terms, resistance management in transgenic plants might include using lower doses of toxin, multiple toxins, and sporadic rather than continual toxin release. Further escape from toxins could be provided in a transgenic crop by designing crops that turn off toxin release at a certain point in a plant's development, or release toxin only in specific tissues rather than throughout the entire plant. In addition, alternating transgenic and conventional plantings might limit the onset of resistance. Physical escape could be provided by pest refuges in which perhaps 10 percent of a field is planted with traditional, nontransgenic varieties.

These are all possible solutions to resistance, but they go against the grain of contemporary agricultural practice, remain theoretical, and are based more on computerized studies than on practical field experience. The few field studies conducted to date indicate that resistance management may be fraught with practical barriers. Further, similar resistance management strategies have been proposed for chemical pesticides, with only limited acceptance by farmers. Thus, transgenic crops could put us on a resistance cycle similar to that induced by our current pesticide management strategies. Nevertheless, a marriage between evolutionary biology and pest management has potential to reduce the onset of resistance to pesticides, and this developing field is well worth watching during the next decade.

❧

The most compelling aspect of the transgenic plant debates has been how the mainstream regulatory process has balanced the opposing poles represented by industry and environmentalists. The two extremes continue to name-call and posture, but it is a rite now without substantive effect, since the battle is essentially over. The regulatory infrastructure has learned its lesson from chemical pesticides and has proven successful at balancing environmental concerns with industry and agricultural needs, although it is doubtful that either side would agree with that statement. Those opposed to transgenic plants claim that regulation has failed, because these bioengineered organisms are making their way into broad commercial use. Industry and farmers complain that it took almost 15 years to move from concept to product approval, much too long from a profit-driven industry perspective or from the point of view of a farmer who sees a considerable portion of his crop and income lost to pests and pest management costs each year.

Regulation has been a success from the public's viewpoint, however. The public has an interest in pest management proceeding, but also in pest management being safe for humans and the environment. The regulatory control of transgenic plants has advanced in careful stages from concept to product that allowed for a thorough evaluation at every step. Products have been released for sale only after convincing evidence that a transgenic crop would perform as advertised without side effects. This is not to say that there are no ecological risks, but they are low, and the probabilities are well-understood. High-risk transgenic crops, even if potentially effective at combating a pest, do not make it through the testing and registration process.

The regulatory procedures for transgenic plants initially were hypercautious, reminiscent of the way moon rocks were treated when they were first brought to Earth by astronauts. There was considerable fear that specimens from the moon would harbor some unknown pathogenic organisms that could devastate nonresistant terrestrial plants or animals, and so the moon rocks were kept in highly confined and isolated chambers until the risk of contamination could be evaluated. Even returning astronauts were kept in medically secure isolation facilities for weeks to ensure that they were not harboring infectious disease originating on the moon.

The caution with bioengineered plants exhibited by regulatory bodies in both the United States and Canada was not driven by moon rocks, however. Old and fresh memories of regulatory mistakes with chemical pesticides motivated the regulators to be more careful with bioengineered plants. The early introductions of chemical pesticides such as DDT and Agent Orange were largely unregulated, and only recently have chemical pesticides been thoroughly evaluated before release. The early nonchalance about pesticide regulation has stiffened, but unfortunately the pesticide bureaucracy has evolved into a labyrinth of overlapping agency jurisdictions, industry lobbying, and approvals of borderline chemicals that fall between the cracks of various regulatory bodies.

The USDA, EPA, and FDA biotechnology regulators did not want to repeat the errors made by their counterparts with chemical pesticides. They developed stringent testing requirements and clearly defined areas of authority right at the beginning of transgenic crop development. Their approach was to favor attracting the wrath of industry with overly cautious rules rather than face the anger of a growing environmental movement. Thus, the first transgenic plants to be tested were treated with considerable caution. Photographs of these early bioengineering field sites show workers dressed in white containment outfits with air-filtering masks, and environmental monitoring devices placed in and around the fields at a higher density than the plants themselves. Experiments were short, and the experimental plots burned with fire and chemicals to destroy any potential threats once experiments were concluded.

As time went on, however, and nothing horrible transpired, regulations for field tests began to relax somewhat, but today remain fairly vigilant. Regulators continue to err on the side of caution, using guidelines stricter than our experience with these plants would suggest was necessary. The first stages of transgenic plant release into the field are still maintained under strict confinement conditions, and these regulations have been effective. The conclusions of the 1989 National Research Council report on transgenic plants would be the same if written today: "Millions of individual plants are tested annually in the United States, and no environmental damage has been documented. Standard confinement practices have been effective at keeping in bounds both poorly domesticated and highly domesticated plants. The committee could document no case of escape of a plant introduction from a confined experimental field plot that has produced problems in natural ecosystems."

The final permission to begin even small-scale commercial testing requires a thorough evaluation of risks before approval. The EPA and the Canadian equivalent Pest Management Regulatory Agency are concerned about the potential of bioengineered plants to themselves become weeds, or for gene-flow to wild relatives to produce hybrid offspring that could become more weedy or invasive. In addition, regulators worry about potential effects of transgenic plants on nontarget species, including humans, and possible impact on biodiversity. Satisfaction of the extensive regulatory standards in place to prevent these problems requires years of research, massive submissions of data, and a lengthy review process. To date, only ten plants bioengineered for pest management have made it to the commercial stage in the United States or Canada, three insect-resistant, six herbicide-tolerant, and one virus-resistant crop, which indicates just how strict the regulatory procedures have been for bioengineered plants in North America.

Perhaps the best evidence that strict regulation is in place to protect us from environmental and health disasters is that nothing of any concern has happened to date—no documented examples of human health problems or deleterious impact of transgenic plants on nontarget organisms in North America, no evidence that gene flow into the weed population has occurred from these transgenic crops, although they have been field-tested since 1986 and commercially available for agricultural use since 1994.

The environmental movement deserves some credit for the original implementation of strict regulatory guidelines from transgenic plants, but it is difficult to understand its members' continued opposition to bioengineered pest management in the face of strict regulations and almost no evidence of harmful side effects. The groups opposed to recombinant organisms began their work in the 1960s and 1970s, when environmental disasters from DDT to Agent Orange, Dioxins to Three Mile Island, were prominent in the news. We have learned from these experiences that close oversight of the scientific and industrial communities is necessary to prevent new scientific and technological advances from becoming harmful. However, transgenic plants no longer appear to be science-gone-mad, and with proper regulation their potential to become environmental Frankensteins is slim. Perhaps it is time for groups opposing bioengineered plants to let go of this issue and move on to other, more substantive problems.

Regulation seems to have found its appropriate level on the issue of transgenic plants. Even so, it is likely that the regulators eventually will err, particularly as large-scale commercialization leads to situations difficult to anticipate in smaller-scale trials. It would be foolish to believe that there will never be environmental damage due to transgenics, but the success rate to date indicates that problems will be rare, and probably acceptable in that severe, permanent impact is not likely.

The one area in which the regulation of agricultural bioengineering seems environmentally problematic is in the commercialization of herbicide-tolerant plants. The USDA, FDA, and EPA are beginning to develop programs designed to reduce the use of chemical pesticides in

United States agriculture as well as other types of pest management, but the approval of herbicide-tolerant crop varieties has the opposite effect, and will substantially increase herbicide use by farmers. The effects of overusing herbicides will be similar to that of insecticides—increased pest resistance and impact on nontarget organisms. The various regulatory bodies have yet to grapple with this issue but rather lean toward approving these crops because there has been no clear direction from Congress not to do so. This obviously is a regulatory area that needs some attention.

The EPA also has expressed the appropriate level of concern about the possibility of pests developing resistance to bioengineered crops and has correctly allowed products on the market even though the agency recognizes the potential for resistance. Nevertheless, product registrations have been made conditional on the development of workable resistance management plans, and the agency has been a partner with industry and academia in sponsoring conferences and working groups to develop resistance management guidelines.

The wisdom of the EPA's resistance management paradigms is being tested today. In 1996 thousands of acres of bioengineered *B.t.* cotton produced by Monsanto were consumed by cotton bollworms, one of the pests that this cotton variety was designed to kill. While it is not yet certain that resistance is behind the bollworm's success, the EPA is considering suspending Monsanto's registration for *B.t.* cotton if evaluations indicate that the insects are becoming resistant to *B.t.* toxins. Clearly, the response to this issue will send a strong signal to industry concerning how serious the EPA is about resistance management, and both environmentalists and industry are watching this case closely.

The final irony in the transgenic debates has been that these products have good potential to provide a means of reducing chemical pesticide use, with the exception of herbicide-tolerant crops. The slow pace at which bioengineered crops have been approved by regulators has taught us that the past and current problems caused by chemical pesticides, such as massive die-offs of fish and birds and extensive reductions in biodiversity, can be avoided with properly regulated transgenic products. Nevertheless, public mistrust of scientists, regulators, and unfamiliar technology has led to vitriolic protests and legal maneuvers designed to prevent bioengineered plants from entering the ecosystem.

Bioengineered, transgenic plants will not solve all pest problems, and it is important to continue regulating these products to exhaustively test their safety prior to release. With appropriate regulation, transgenic crops can replace some conventional, chemically based agricultural pest management, and differ from other alternative technologies in one major respect: of all the alternatives to chemical pesticides, bioengineering is the only one that has caught the attention of industry in a substantial way. If alternatives are to become mainstream, they must in the end be marketable, and industry has clearly made its judgment that other options like pheromones, parasites, and predators are fringe players in the pest management arena, at least for the foreseeable future.

There appears to be little risk to trying most bio-engineered plants, considerable potential to reduce the use of chemical pesticides, strong industry and agricultural support, and a careful regulatory process in place to oversee transgenic crops. Critics of transgenic plants might consider one point: even the worst-case scenario of occasional problems with bioengineered plants might be preferable to spraying billions of pounds of chemical pesticides into our world each year.

CHAPTER NINE

Moving Beyond Rachel Carson

"A truly extraordinary variety of alternatives to the chemical control of insects is available. Some are already in use and have achieved brilliant success. Others are in the stage of laboratory testing. Still others are little more than ideas in the minds of imaginative scientists, waiting for the opportunity to put them to the test."

Rachel Carson, *Silent Spring* (1962)

Rachel Carson's *Silent Spring* was not a complicated book. It had two simple take-home messages of significance for pest management. First, chemical pesticides can be dangerous to ourselves and the environment and should be the last rather than the first technique used to control pests. Second, there are biologically based alternatives to synthetic pesticides, and these ecologically friendly methods should be studied and implemented.

These messages, which were controversial in her time, today have become motherhood and apple pie to the pest management industry. I doubt that there is a single pest manager anywhere in North America, and probably in the world, who would disagree. Prevailing public opinion clearly favors responsible approaches to environmental issues as well. Yet, toxic chemicals remain the first line of defense against pests, and synthetic chemical pesticides the overwhelmingly dominant method of structural, forest, home, urban, and agricultural pest control. Alternatives remain just that—fringe techniques occasionally used, token attempts that barely dent the enormous mass of chemical pesticides applied each year against pests.

Each of the preceding eight chapters tells a particular story in which various alternative methods of pest control have been attempted. From insect-killing bacteria to pheromones, parasites and predators to sterile insects, integrated pest management to transgenic plants, these techniques have failed to replace our pesticide-dependent control paradigms. Each of these methods or systems is scientifically interesting, but its implementation has been limited or prevented by public opinion, economic considerations, industry resistance, government regulation, poor communication by professionals to the public, and ultimately our sense that we must dominate pests rather than manage them. Taken together, these stories reveal the disappointing reality that biologically based technologies for pest control have yet to enter the mainstream of modern pest management, and remain mostly in the realm of the possible rather than the actual.

The history of synthetic chemical pesticides and their biological alternatives is fraught with ironies. Consumers worry about the effects of agricultural pesticides on the natural environment, but in their own homes and gardens they use highly toxic chemicals against fairly harmless pests, demanding fast action with zero tolerance. City residents howl in protest when pest managers propose spraying the environmentally friendly bacteria *Bacillus thuringiensis* to control gypsy moths. Then in other situations, the public backs costly alternative programs with their tax dollars, such as Sterile Insect Release for codling moth, only to find that the scientific and economic rationales are weak. Industry avoids some biological technologies such as pheromones, parasites, and predators but invests heavily in others like transgenic plants, to the accompanying outcry of a public with a nonspecific fear of genetic manipulation. Industrial and scientific claims for the safety of a new method such as bioengineered plants are met with scorn and suspicion by environmental groups and the media, because of the poor credibility established by earlier generations of industrial and academic representatives who claimed human and environmental safety for now-banned or highly restricted chemical pesticides.

The ultimate irony may be that the very agency established to protect us from chemical pesticides has wound up inadvertently encouraging them, and discouraging alternatives. Well-intentioned protests against the toxic pesticides and other pollutants used in the 1950s and 1960s led to the creation in 1970 of the U.S. Environmental Protection Agency (EPA), following a decade of intense legal and political pressure from citizens' groups like the Environmental Defense Fund, Friends of the Earth, Greenpeace, and the Conservation Law Foundation. These organizations lobbied hard for legislation to control pesticide use through a strengthened Federal Insecticide, Fungicide, and Rodenticide Act (FIFRA), and also for the establishment of a government regulatory body that would act as a balance to the Department of Agriculture, whose attitudes toward pesticide use at that time can best be described as unenlightened.

The EPA first acted by banning the use of DDT in 1972 and by setting up a rigorous system to evaluate new pesticides and regularly reevaluate registered chemicals. Today, the EPA's Office of Pesticide Programs (OPP) has an $87 million annual budget, 760 employees, and a staggering workload to oversee a thousand current or proposed active ingredients formulated in well over 20,000 different products. Even with this enormous staff and budget, the OPP is facing an incredible backlog of decisions. For example, the OPP has been told by Congress to review the registrations of older pesticides, a process that is occurring at a rate of about 35 per year. With 860 registered pesticides to date, and only 121 reviewed by the end of 1995, it will take well into the year 2016 to even review old chemicals, let alone catch up with licensing new ones.

Unfortunately, the EPA has never advanced far beyond its original mandate to protect us from the harmful side effects of the most poisonous chemicals. The narrow EPA focus on extremely toxic compounds rather than a broader focus on fostering well-balanced pest management has increased our dependence on chemical pest control and discouraged the development of alternatives, because so many chemicals do eventually make their way through the system and get licensed for use.

The underlying pest management philosophy of the EPA is reflected in the mission statement of the Office of Pesticide Programs. The main mandate of the OPP is "safeguarding public health and the environment from pesticide risks, and ensuring that pesticides are regulated fairly and efficiently." Although this is obviously important in a chemically dependent system, it is noteworthy that the OPP's prime directive is not something like "the reduction of pesticide use, and the encouragement of  safer, biologically based technologies." The administrative focus on regulatory protection from pesticides rather than the development of alternatives is a significant reason that chemical use is so widespread in pest management. The overwhelming number and toxicity of chemical pesticides has forced us to react by constructing an extensive, ponderous bureaucracy to regulate their use. The regulation of synthetic chemicals is necessary, but our need to oversee pesticide use has overwhelmed the pest management agenda, leaving reduced government resources and little energy or imagination to explore and implement alternatives.

Perhaps it should not be surprising that our regulatory bodies have been kept busy with pesticide regulation rather than reduction. Congressional legislation to date has demanded an unrealistically low level of pesticide tolerance, driven largely by consumer fears of cancers induced by pesticide residues in food. The major legislation that defined how the regulatory process would work in practice was the 1958 Delaney Clause amendment to the Food, Drug, and Cosmetics Act, which stated, "No additive shall be deemed safe if it is found to induce cancer when ingested by man or animal." This statute has been interpreted to mean that there basically are no safe levels of exposure, and in response, the EPA and FDA have erected ponderous bureaucracies to ensure that there are virtually no pesticides in food, a level of vigilance that has controlled the regulatory agenda since the Delaney Clause was passed.

The extreme caution exercised by regulators was evident early on, in the cranberry scare of 1959. Just before Thanksgiving of that year, Arthur Flemming, Secretary of the Department of Health, Education, and Welfare, announced that cranberries from the western United States had residues of the weed killer aminotriazole (3-AT), a substance that was thought to induce thyroid cancer in rats when fed at high doses. In the ensuing panic, the U.S. government seized 3 million pounds of cranberries and growers lost an estimated $40 million in sales. Although the facts gradually emerged that most U.S. cranberries did not contain residues of the weed killer, and that in any case it took extremely high doses of 3-AT fed continuously to rats to induce cancerous growths, the perception that any pesticide residue in food would cause cancer became indelibly established in the public lore.

The same scenario has been repeated numerous times, most recently with Alar on apples in 1989. Alar is a growth regulator sprayed on apples to prevent premature drop of ripening fruit from trees. It was brought to the attention of Americans on February 26 of that year in a CBS television program that proclaimed it to be "the most potent cancer-causing chemical in our food supply." In the ensuing panic, growers in Washington State alone lost at least $125 million in sales. As with cranberries, however, Alar causes cancer only at doses much higher than those found as residues in apples. The levels found in apples were and continue to be considered safe by most regulators and scientists involved in carcinogenicity studies.

The key issue in these debates has been whether we can tolerate a vanishingly small cancer risk of pesticide residues in food, such as a statistical probability of one in a million, or whether that risk needs to be reduced to zero. This debate in recent years has been strongly influenced by Bruce Ames, a biochemist at the University of California, Berkeley, who has argued that most foods contain carcinogens naturally, and at higher doses than pesticide residues. He noted in a 1991 paper that "almost every fruit and vegetable in the supermarket contains natural plant pesticides that are rodent carcinogens. The levels of these rodent carcinogens . . . are commonly thousands of times higher than the levels of synthetic pesticides." He went on to point out that the FDA calculated the average U.S. intake of synthetic pesticides in food as 0.09 mg per day, compared to 1.5 g of naturally occurring and equally carcinogenic substances found in the diverse food products we consume.

In 1996 Congress finally addressed this regulatory problem with bill H.R. 1627 that replaced the zero-risk Delaney Clause with a new standard providing for "a reasonable certainty that no harm will result from the aggregate exposure to the pesticides' chemical residue." In addition, the bill requires the EPA to accelerate its review of older pesticides, thereby reducing risks from earlier generations of pesticides now licensed for use that were not tested as extensively as more current compounds. This amendment generally has been viewed favorably by environmentalists, because it allows for a reasonable use of pesticides while preventing misuse or overuse that might leave unacceptable residues in food. In addition, it may finally free the EPA from the burden of excessive regulation, and allow for more resources to be directed toward pesticide reduction programs that in the end will do more to protect consumers than enforcement of the Delaney Clause did.

❧

The EPA has realized that a better approach to pesticide regulation might be to reduce pesticide use, leaving fewer chemicals to regulate and fewer uses to oversee. They recently began a new program called Biopesticide, Risk Reduction, and Reinvention Initiatives, designed to encourage "the introduction of a new generation of biological pesticides, reduce pesticide risks through environmental stewardship, and reinvent OPP organizations." Their flagship project is called the Pesticide Environmental Stewardship Program, described by EPA administrator Carol Browner in an attractive brochure soliciting public and commercial partners for this project: "Voluntary pollution prevention has been a cornerstone of our efforts to protect human health and the environment, and this new pesticide partnership is an important step toward that goal."

The EPA is a rather late entrant in the pesticide reduction field. Other government organizations have been attempting for many years to stimulate biologically based technologies through research, though their success rate has been less than robust. Government efforts to encourage alternative methods of pest management were summarized in *Biologically Based Technologies for Pest Control,* put out by the Congressional Office of Technology Assessment in 1995. Perhaps the most astounding finding in this report is that total federal funding for research in U.S. government laboratories involving biologically based technologies (excluding bioengineering) has been over $100 million a year for the last ten years. In 1995 the federal government invested $165 million into research involving pheromones, microbial pesticides such as *Bacillus thuringiensis,* plant pathogens, biological control using predators and parasites, and other types of biological methods. The largest chunk of this support was directed to the Agricultural Research Service of the U.S.

Department of Agriculture, $104 million, with about 1,200 employees devoted to 300 individual projects.

What is perplexing about these numbers is that the market value for sales of all biological products was only $100 million in 1993, the last year for which complete data are available, with over half of those sales coming from a single product, *B.t.* That is, in spite of a staggering investment of $1.3 billion in federal research dollars spent on alternative technologies over a 10-year period, the commercial impact of these technologies is still only 1 percent of the $8.5 billion annual economic value of synthetic chemical pesticide sales. Clearly, this research has not provided good bang for the buck.

The OTA report did not mince words about why research in federal laboratories has failed to have a commercial impact in pest management proportional to the dollars invested:

> Despite its size, this expenditure of federal funds for research appears to be largely uncoordinated and to lack adequate prioritization. Widespread agreement exists that basic research on biologically based technologies is poorly linked to on-the-ground applications. One reason is a lack of research necessary to translate findings into practical field applications, in part because no federal research agency takes responsibility for this function . . . Use of biologically based technologies requires a significant level of information and knowledge, and farmers often lack clear-cut instructions or authoritative sources of advice on how to apply them, and most extension agents have had little if any formal exposure to biologically based approaches.

While it is tempting and convenient to blame government for the failure of these new technologies to excite the marketplace, the poor commercial success of alternative technologies has many additional causes. The broad professional pest management community, including university and government researchers, extension agents, and private consultants, also have been ineffective at translating research findings into management practice.

The research itself has been impressive. For example, the annual meeting of my own professional society, the Entomological Society of America, is attended by a few thousand entomologists presenting about 1,500 papers, almost all of them relevant in some way to biologically based technologies. These days, very little of the work presented has to do with synthetic chemical pesticides. However, in the twenty years I have been attending these conventions, I have never seen more than three or four booths in the commercial display area that represent any type of biologically based control technology that is available for sale to farmers or other pest managers.

It is difficult to assess the role of scientific research in the implementation of alternative technologies without falling into one of the opposing extremes of academic-bashing or unabashed promotion. Both extremes are prevalent in pest management, with researchers using exaggerated language in grant applications to extol the potential of their latest biologically based ideas, and on-the-ground pest managers wondering what happened to the previous batch of interesting-sounding alternatives. The reality usually lies in the middle. Research has contributed to pest management, and many research projects have led to real-world applications. Nevertheless, the overwhelming fact that we use over a billion pounds of pesticides each year in the United States, a number that increases annually, indicates just how far from success the research community has been in developing commercially viable rather than just potential alternatives.

Integrated Pest Management exemplifies this failure. IPM is a holistic philosophy more than a technique, and differs from earlier pest management paradigms in not attempting to eradicate pests. Rather, the attitude of IPM practitioners is to apply the minimal level of control that will maintain a pest below an economic damage threshold. IPM also differs in using multiple techniques rather than following only one pest management strategy, and also in its inherent goals of reducing pesticide use and minimizing damage to nontarget organisms and ecosystems. IPM recognizes the complexity of biotic interactions in the field, and in theory appears to supply the ideal philosophical underpinning for biologically based technologies.

Unfortunately, that complexity has been its downfall in practice. Academic researchers may prefer to deal with complex issues, but farmers

require simpler solutions. IPM is not simple; it requires a major commitment to gathering knowledge about a pest's biology, and a major investment in the extensive training necessary to be an effective IPM practitioner. Setting up a proper IPM program includes understanding the biology of the crop or resource, detecting key pests, knowing their biology, recognizing the kind of damage pests inflict, initiating studies on their economic impact, identifying environmental factors that influence pest survival, perceiving concepts and methods that individually or in combination could help to suppress pest populations, identifying weak points in pest biology, narrowly focusing control practices at these weak links, practicing continuous monitoring, developing and applying methods that preserve the ecosystem, and many, many more.

IPM is realistic in emphasizing the complexity and subtlety inherent in pest systems, but it is not directly approachable for most farmers, exterminators, and other pest managers because of the level of expertise and increased costs required to practice this craft properly. Even highly trained extension agents may have considerable difficulty in advising about the complicated models and methods often necessary to implement IPM.

Integrated Pest Management has had some impact, especially by encouraging pest monitoring prior to chemical treatments. Pesticide applications today usually are recommended only when pest populations increase beyond an economic damage threshold rather than on a scheduled basis. However, the early promise of IPM has given way to pessimism even among its proponents. The cynical term "Integrated *Pesticide* Management" in vogue among pest managers today is evidence of how far IPM has fallen in their esteem. Many pest managers pay lip service to the inclusion of IPM as part of their management, but the IPM performed is often an ersatz, pesticide-driven version, rather than the complex, ecologically based program originally proposed by the research community. IPM may tell a pest manager when to treat, but the treatments still are overwhelmingly chemical.

William Olkowski, an editor of *The IPM Practitioner,* in 1991 described the problem of pesticide dependence and the warping of IPM philosophy this way:

> As the IPM concept has become more well known, the phrase alone

> has been increasingly identified by the public as "good" due to the assumption that IPM is synonymous with least-toxic pest management. As a result, pest managers and institutions have eagerly jumped on the bandwagon, claiming that IPM is what we have been doing all along . . . It is critical that the mere term "IPM" not be adopted without its content. For example, it is a misunderstanding to call rotation of pesticides "IPM" simply because more than one material is used.

Linda Edwards, the Okanagan Valley pest management advisor I spoke with about codling moths, is one practitioner of IPM who feels that the concept has become almost meaningless. Her company, Integrated Crop Management, was commissioned to compile a directory of Canadian expertise in IPM, a list that eventually included about 700 individuals and 200 organizations claiming to practice IPM. The reality, however, is that those faced most immediately by a decision about whether to use a pesticide do not have the luxury of IPM philosophy but rather have to face economic reality. As Edwards put it, "If there's one word that would sum it up, it's economics. It's fine to say to someone 'Don't use pesticides,' but try to hand-weed economically. It's easy to sit on the outside and say 'Why use Roundup?' but the picture is often very different on the inside. At the grower level, we're at the bottom of the food chain, the end of the line. All those people who tell us how to do it, they don't have to walk the walk."

❧

The tendency of the research community to discover scientifically interesting but impractical alternatives to solve pest management problems may be one contributing factor to the continued high level of chemical pesticide use, but economic issues are at the heart of our dependence on pesticides. The difficulty with pesticide economics is that costs and benefits are calculated by users on short-term time frames, but real costs and benefits may be more accurately portrayed by considering the hidden costs to society above the purchase and application price for chemical pesticides.

Tobacco provides a good analogy, particularly since one of its most toxic ingredients is the insecticide nicotine. Cigarettes are cheap to the smoker, even with taxes, but their real costs are considerably higher than their purchase price. Smoking adds enormous indirect costs for health care, fire damage, construction of ventilation systems, and other bills that society as a whole must pay. These costs can equal or even exceed the purchase price for each pack of cigarettes.

Similarly, pesticides are relatively cheap for users, but their up-front purchase price of $8.5 billion per year is doubled by hidden costs to society of $8 billion annually to pay for pesticide-induced medical problems, environmental clean-up, government regulatory costs, and losses of beneficial organisms. If these real costs were calculated into the purchase price, then pesticide costs would double, and their apparent cost-effectiveness would be considerably diminished. Unfortunately, the marketplace works more simplistically, focusing on immediate expenses and annual profits, and these long-term societal-level costs do not get added in to the purchase price.

Another way to consider pesticide economics is to look at the relative costs and benefits of reduced pesticide use and the increased use of alternative technologies. David Pimentel and Hugh Lehman, in their 1993 book *The Pesticide Question*, compared the cost for farm use of pesticides in the United States to the costs that would result from using biologically based alternatives to replace some pesticide use. In one example, they calculated that control of corn insect pests in the United States would cost $280 million a year if farmers used alternative methods such as crop rotation, planting of resistant varieties, and poison stations baited with attractant pheromones. This $280 million price tag works out to $60 per acre treated and is considerably higher than the current $187 million, or $36 per acre, spent on purely insecticidal control. These calculations would not economically justify the use of biologically based methods to a corn farmer trying to earn a living in today's highly competitive agricultural environment. However, their proposed pest management methods would reduce insecticide use by 86 percent, an attractive proposition if our society were sufficiently concerned with nontarget impact of chemical pesticides.

Pimentel and Lehman did similar analyses for a wide variety of field, vegetable, fruit, and nut crops and determined that alternative technologies supplemented by occasional applications of chemicals could easily reduce overall pesticide use by 50 percent. The total cost to farmers for this pesticide-reduced management program would be $818 million, a 20 percent increase in direct costs for pest control but only a 0.5 percent increase in total food production costs. Retail prices paid by consumers would increase about 1.5 percent on average according to their calculations, but we would save about $450 million in side effect costs due to reduced pesticide impact on human health and the environment.

Even if their calculated costs for alternatives were two or three times too low, their approach would still suggest that the costs for pesticide reduction and implementation of alternative controls are not beyond reach if our society decided that pesticide reduction was a major objective in pest management. Clearly, however, short-term user economics predominate in pest management decision-making. Since direct, out-of-pocket costs for pesticide-based control are cheaper than biologically based methods, it would take a major and probably legislated imposition of alternative technology by government to induce any substantive movement toward a pesticide-reduced system.

The higher costs of biologically based technologies were apparent in almost every example I could find. Pheromone-based mating disruption for codling moth costs $85 an acre more than insecticide sprays. Biological control with parasites and predators is only cost-effective in the closed universe of the highly lucrative greenhouse industry, where management costs of up to $5,000 an acre are considered affordable. Structural pest controllers charge $30–45 an hour more to utilize Integrated Pest Management techniques compared with hosing down a building with pesticides.

The end users of pesticides, such as farmers, homeowners, structural pest control operators, and foresters, have by and large chosen the lower costs of pesticides over the more expensive but environmentally safer alternatives. Biological products reside mostly in small niche markets and barely register sales in the pest control marketplace compared with chemicals. These alternative technologies do have some appeal for those few consumers willing to pay more for environmental safety, or who recognize that the side effects of pesticides reflect real rather than actuarial costs. These markets to date have been minuscule, however, if sales volume accurately reflects consumer acceptance.

Like their customers, industry has been reluctant to embrace biologically based technologies, and has directed relatively little investment in this area compared with chemical pesticides. In some ways this is surprising, because alternative technologies are much easier to bring to market than chemicals. The EPA has streamlined the regulatory process for biologicals so that required tests and fees for a typical registration application cost only $200,000. An average chemical pesticide requires a tremendous amount of safety and efficacy testing for registration, at a typical cost of about $20 million. Research, development, and registration for a biological method takes roughly three years and costs $1–2 million, compared with eight to ten years and $25–80 million for a chemical pesticide.

However, chemical pesticides are considered to have so many commercial advantages over biologically based technologies that the inexpensive and rapid registration and testing process for biologicals is not enough to attract much industrial interest. For example, biologicals generally are effective against one or a very few pests, whereas chemicals tend to be broad-spectrum. Most biological methods are sensitive to weather, whereas pesticides tend to work well under a broader range of environmental conditions. Also, most alternative technologies have short shelf lives and field persistence, while chemical pesticides can be stored for years and may persist for days, weeks, or even longer after application. Most significantly, a pesticide producer can expect revenues in the hundreds of millions of dollars for a successful pesticide, while the track record for most biologicals at best shows extremely modest profits, and more frequently losses.

Transgenic plants and to a lesser extent the microbial pesticide *Bacillus thuringiensis* are two interesting exceptions to industrial and consumer preferences for chemical pesticides over alternatives. While transgenic products are only just reaching the commercial stage, current and projected costs suggest that bioengineered crops will control pests for prices similar to, or lower than, those of pesticides. The favorable economics and efficacy of transgenic plants has generated considerable industrial and consumer interest in this technology. For *B.t.*, the costs are only slightly higher than chemical pesticides, and its record of fairly good activity against pests combined with its excellent health and environmental safety record have made *B.t.* a profitable commercial product. It is therefore not surprising that industry has embraced research and development of transgenic products and *B.t.*, while alternative technologies such as pheromones, biological control, and other more costly microbial pesticides languish at the edges of commercial pest management.

It also is interesting that the two alternative approaches most attractive to industry and pest managers—transgenic plants and *Bacillus thuringiensis*—are the very ones viewed with the most alarm by the public. Indeed, a significant, almost overwhelming, problem we have had in moving beyond Rachel Carson is that the public is highly mistrustful of new technologies in pest management, especially when they are embraced by industry or government. This lack of trust has a historical basis, but the reaction of the public and media to new pest management technologies has been out of touch with reality in issue after issue. In some cases the public and journalists accept pest management uncritically where considerable skepticism would be warranted. In others, they are highly critical to the point of panic, where the method being used is safe, effective, and well-tested.

Take structural pest extermination and codling moth Sterile Insect Release as two examples. While these two areas of pest management appear to have little in common, they share the mutual distinction of enjoying public acceptance where skepticism is warranted. We often tolerate sprays of noxious pesticides in our homes, backyards, and workplaces with hardly a word of protest, and certainly with little understanding of whether the spray is safe or even necessary. A costly SIR project against codling moths that is environmentally friendly but was oversold in its expectation of eradication was similarly accepted uncritically by consumers desperate for a solution. In both situations, the lack of public outcry is difficult to fathom.

Public reactions to the *B.t.* spray in Vancouver against gypsy moths and to bioengineered plants also have shown little relationship to reality, but in the opposite direction. For *B.t.*, a handful of protesters fueled by a scandal-seeking press held the city's attention for months on a non-issue. *Bacillus thuringiensis* is safe; it has been used for more than 35 years in countless spray programs against insects, and in that time has not generated a single medical or environmental incident of any consequence. Bioengineered plants also generated considerable and perhaps appropriate public concern when first proposed, and as a result potential bioengineered products have been exhaustively tested and are rigorously regulated. Yet today, when the few examples of transgenic plants making it through the regulatory and registration process have little possibility of causing damage, protests continue. Scientists, industry, and the government fulfilled a societal mandate with both *Bacillus thuringiensis* and transgenic plants to develop safe alternatives to chemical pesticides, yet were met with an excessive storm of public protest when these new technologies hit our cities and fields.

Public and media attention to pest management, now as in the past, continues to be important in ensuring that new technologies are safe. But when concern enters the realm of the irrational, the public interest is not being served. Our society seems to have equivalent phobias about pests and pest management, and we exhibit just as little ability to discriminate between a real and an imagined pest as we do to discern differences between risky and safe pesticide use and effective or ineffective alternative technologies. As a result, our choices about which pests to control and what management techniques to use to control them are often inconsistent with our professed interest in lowering health risks to humans and adverse impact on the environment.

Pop culture has played a significant role in fostering pest phobias. To boomers who grew up in the 1950s, movies such as *The Fly* portrayed a suave inventor turning into a gross and disgusting housefly, yet in reality the common house fly is harmless. Another movie, *The Swarm*, depicted an immense swarm of bees destroying a nuclear weapons launching facility in Texas, as well as the city of Houston, yet bees are beneficial because they produce honey and pollinate crops and natural vegetation. *Arachnophobia* terrorized us with spiders, but most spiders are useful biological control organisms because they catch and eat pest insects.

Our Hollywood-enhanced pest phobias have resulted in a considerable amount of pest control being directed against pests that pose no risk to our health or our food supply. For example, a surprising level of pesticide use is focused on maintaining consumer-driven cosmetic standards for food. David Pimentel and coauthors described this problem in *The Pesticide Question:*

> The American marketplace features nearly perfect fruits and vegetables. Gone are the apples with an occasional blemish, a slightly russeted orange, or fresh spinach with a leaf miner . . . Wholesalers, processors, and retailers have been increasing their cosmetic standards for various reasons, including perceived consumer demand. The results have been higher economic costs for pest control, widespread environmental and human health problems caused by pesticides, as well as higher contamination levels of insecticides and miticides in fruits and vegetables.

Much of this cosmetic pest control is directed against insects found on food, whether they damage the crop or not. Standards for the numbers of insects found in grocery produce are set by the Food and Drug Administration. The allowable numbers of aphids and leaf miners in spinach have dropped in the last sixty years from 110 aphids per 100 grams of spinach to only 50, and for leaf miners from 40 to 8 per 100 grams, because of consumer demands for produce that is largely insect-free. Neither aphids or leaf miners present a threat to human health, yet pesticide use on spinach has increased from one or two to three to six annual treatments to achieve the more stringent insect-free requirements. The squeamish consumer could easily wash these insects from produce, but we have chosen to kill them with pesticides rather than have to face them in our sinks.

Similarly, surface blemishes are common from insect and mite pests that do not otherwise enter or harm fruit. For example, citrus rust mites are only found on the outside of oranges, and cause external russeting or bronzing. Although they do not affect nutritional quality, flavor, shelf life, or any aspect of oranges other than their external appearance, consumers demand blemish-free fruit. Thus, 80 percent of the Florida citrus crop is sprayed an average of three times a season just to prevent discoloring of orange skins.

In total, 60 to 80 percent of pesticide use on oranges, 40 to 60 percent on tomatoes, and 10 to 20 percent on most other fruits and vegetables is strictly for cosmetic reasons. The choice has been ours; each time we turn away a head of lettuce because we see an insect on it, or an apple with an external blemish, we are encouraging the use of more pesticides. Virtually every consumer would claim to desire pesticide-free food, yet consumer pest phobias are a major factor encouraging increased pesticide use.

Biologically based technologies often work well enough to control pests at a reasonable level but may not reach the level of pest control required by a pestaphobic public. A *Bacillus thuringiensis* spray would kill most spinach leaf miners but not enough to satisfy insect-fearing consumers. Biological control organisms such as the predatory mite *persimilis* would eat many citrus rust mites, but not munch a sufficient number of them to prevent all russeting. Attractant-based cockroach traps can catch most cockroaches, but only pesticides can quickly kill them all. We demand total pest control, and the price we pay for that perfection is a billion pounds of pesticides sprayed each year in our homes and on our fields.

A final perplexing and ironic aspect of the pesticide treadmill is that pesticides are becoming less and less effective as we use more and more of them, yet we seem to be blocked by inertia from reducing pesticide use. Resistance to pesticides has forced us continually to invent new chemicals that quickly lose effectiveness. The percentage of crops lost to pests has increased from 31 percent in the 1940s to 34 percent in the 1950s to 37 percent in the 1980s, yet we continue to respond by spraying more chemicals rather than developing smarter control techniques.

Our progress since *Silent Spring* in learning how to manage pests in environmentally friendly ways has been mixed at best. We have developed an encompassing regulatory bureaucracy to protect ourselves against the worst of our pesticidal poisons, but it is a regulatory system that is ponderous rather than progressive, reactive rather than preventive, and ineffective at reducing pesticide use. Our scientific community has exhibited wonderful creativity at inventing new pest management concepts but has been less than capable at bringing these ideas to commercial fruition. Industry pays some attention to alternatives, but biologically based technologies are simply not economically attractive to the major player in the pest management industry. The public, highly aware of potential environmental pollutants, has demonstrated a serious lack of discriminatory ability in acting on what is known about effectiveness and danger, usefulness and safety.

I think Rachel Carson would have been disappointed. Her vision of pest control was an ecological one, in which we would cleverly use our understanding of pest ecology and behavior instead of chemicals to manage rather than eradicate pests. She wrote in *Silent Spring*, "We should no longer accept the counsel of those who tell us that we must fill our world with poisonous chemicals; we should look about and see what other course is open to us." Carson looked about and saw a potential world of ecologically based alternatives, including enhancement of natural predators and parasites, introduction of reared ones, sterile insect release, pheromonally based control methods, microbial diseases of pests, and growth regulators that act to disrupt highly specific insect or plant metabolic pathways important for development.

Her confidence in the potential of biologically based alternatives has not been realized. Any one of the factors that have encouraged pesticide use and discouraged alternatives might not on its own have led us so far down the chemical path, but together they have defined pest management as overwhelmingly chemical. Economics, public ignorance, regulation, habit, inertia—whatever the reasons, we clearly are not where Rachel Carson would have liked us to be today. We have realized Carson's worst nightmare rather than her greatest hope.

If Carson made an error in *Silent Spring*, it was in underestimating the difficulty of managing pests in our modern world. She did not adequately appreciate the power of our phobias, the imperative of short-term economics, the complexity of interactions not just among plants and animals but within our species' own society that would prevent biologically based technologies from becoming management realities.

She concluded *Silent Spring* by writing that "the control of nature is a phrase conceived in arrogance, born of the Neanderthal age of biology and philosophy, when it was supposed that nature exists for the convenience of man." Government, agricultural, academic, or industrial organizations have not imposed pesticides on us; it has been our own attitudes about nature and management that ultimately have led us down a chemically dependent road that few of us would have consciously chosen. Until we change those attitudes, pest management will continue to be based on synthetic pesticides.

By more realistically defining what is a pest, and choosing to deal only with organisms that truly cause us harm, by deciding to manage rather than eradicate, and acting against pests only when the danger they present becomes economically or medically significant, and by choosing methods that are the least damaging to our own health and to the environment, even if they appear more costly, we can reverse the trend toward ever more toxic and ineffective methods of pest management. We possess the unusual ability among animals to assess the implications of our own actions, and we can choose the long-term benefits of biologically based technologies over the short-term economic advantages presented by chemical pesticides. We are capable of directing and funding our scientific establishment more wisely, to move beyond interesting concepts to focus on implementation and commercialization of alternative technologies. As individuals and in our collective media, we can be more responsible about which methods we use and advocate and which ones we ignore or protest against.

We must also get our metaphors in order. In the past, we have viewed pests as the enemy and pest control as a military problem, but we will never win the "war" against pests. Our gravest mistake has been in setting a pest agenda that considers pests a problem that must be controlled rather than an integral part of nature that we should manage in effective, environmentally responsible ways. Ironically, by waging wars on pests that we cannot win, we have turned ourselves into nature's losers. More is at stake here than human and environmental health, although both would benefit significantly from reductions in pesticide use. The most important effect of our chosen role in nature as dominators rather than stewards has been to lose our sense of place in the biological world around us. In the end, this may be the steepest price we have paid for failing to heed the message of *Silent Spring*.

CHAPTER TEN

A New Pest Ethic

"We enjoy the fruits of the plains and of the mountains, the rivers and the lakes are ours, we sow corn, we plant trees, we fertilize the soil by irrigation, we confine the rivers and straighten or divert their courses. By means of our hands we create as it were a second world within the world of nature."

Cicero, *De Natura Deorum II* (44 B.C.)

Our most significant failure in pest management is that we have not worked from a unifying concept, a bedrock objective to govern how we manage pests. Because we lack an encompassing ethic, we spray. We must deal with pests, and only the most militant environmental extremists would argue against managing harmful organisms. However, what we consider a pest, and how we deal with those pests, is not a given. We can decide the rules of engagement, define the line where our human duty to protect ourselves ends and our obligation to nature begins.

To date, that line has been drawn with chemical pesticides, a weapon that simultaneously protects us by killing pests but also injures us with side effects that can harm our health and that of the environment around us. Pesticide use has been out of balance, and alternative technologies overwhelmed by the chemical campaign we wage against pests. We continue to use chemicals to a much greater extent than necessary for pest control, as the first line of defense against pests rather than the weapon of last resort.

A viewer from space scanning the Earth would see chemical pesticides permeating our homes, fields, and forests, and perhaps would assume that chemically based pest control was a paradigm that our society had agreed upon, a carefully considered decision that this was how we wanted to control pests. Not so; none of us has decided that pest management should be chemically defined. If we were to make a deliberate choice, I am confident that we would choose to reduce pesticide use.

Unfortunately, reducing pesticide use is not an easy objective to accomplish. Pest management is complex, mediated by an interlocking set of social, political, economic, scientific, and personal factors that make it difficult to modify the current chemically weighted system. The simpler era when DDT was clearly evil, and banning DDT an obvious good, has long since passed. Today, we have banished the most poisonous substances, but this has not solved our problems with pesticides. Instead, we simply overuse more of the moderately toxic.

We can do better, beginning with the development of an ethic to direct future pest management decisions away from chemicals and toward biological methods. Pesticide use is less crucial, and alternatives more possible, than our current pest management practices would suggest. We can begin the transition to a biological paradigm for pest management by formulating a set of simple principles, a pest management ethic:

(1) Chemical pesticides should be the last method used for pest control, not the first.
(2) Pest control should aim to manage pests, not eradicate them.
(3) Only pests doing substantial damage should be managed, and only when their damage approaches an economically significant threshold.

Of course, it is easy to propose idealistic principles for environmentally friendly pest management, but these principles will be adhered to only if there are practical ways to implement them. A pragmatic place to begin is to consider what would happen if we managed pests with less pesticide. Would we be successful, and could we afford it? Decreased food production and increased human health problems are two potential consequences of reduced pesticide use, but both are unlikely. Independent studies by both the Congressional Office of Technology Assessment and the National Academy of Sciences indicate that it is possible to maintain crop yields at current levels while using half the amount of agricultural pesticides; reductions in pesticide use of greater than 50 percent might lead to reduced yields. As we have seen, the pesticide reduction program implemented by the U.S. General Services Administration was effective in almost eliminating pesticide use in over 100 large government buildings, with no increase in pest damage. Alternative techniques can work to reduce pesticide use on our farms and in and around our offices, homes, and industries.

But can we afford the prices associated with alternative methods? In a few cases, biologically based methods are less expensive than chemical pesticides, such as pest management using *Bacillus thuringiensis* genes in transgenic plants. In most situations, alternative pest management is more expensive, but the costs of decreased pesticide use are still small relative to overall management costs. For example, the cost of a 50 percent reduction in agricultural chemical use would be an increase of only 1.5 percent in the price of food at the grocery store; and while an Integrated Pest Management approach to structural pests might cost twice that of a chemically based program, that cost is still minor relative to total building management expenses.

The success of many European countries in accomplishing a 50 percent pesticide reduction program also suggests that pesticide reduction is practical. Sweden, Denmark, and the Netherlands all have reduced pesticide use without serious economic consequences or outbreaks of pest-borne diseases. Their successful programs involved changes such as higher tolerance for certain pests, lower doses of chemical pesticides, better-calibrated spray equipment, less frequent sprays, crop rotation, improved sanitation in food packaging and serving establishments, better building design and construction to exclude pests, and an increase in the use of biologically based technologies against pests.

The experience of the Canadian province of Ontario is also instructive. According to Gordon Surgeoner and Wayne Roberts, its beginnings were less than ideal, but the program was eventually successful, nevertheless:

It would be a triumph of virtue if one could say that the program to reduce pesticides by 50 per cent in the Province of Ontario was based on a consultative process between farmers, agriculture researchers, politicians, and the general public. As in many endeavors the truth is more revealing. In the summer of 1987, there was a provincial election, and various political parties were seeking issues on which the public could make decisions. Public-opinion polls in Ontario had indicated that the public in general was concerned about pesticides from both a human health and environmental perspective. Quite simply, the party which won the election had as one of their platforms a promise to reduce pesticides by 50 per cent.

The Ontario election certainly was not decided only by this issue, and considerable work had gone into this new pest management vision before it got caught up in the political process. Surgeoner and Roberts add that "prior to the election, there had been an ad hoc committee of pest-management specialists including provincial, federal, and university personnel who had attempted to determine where the province should be going relative to pest management . . . In essence, the political platform would not have developed without a preliminary agenda having been developed by pest management personnel. The public would have shown little interest in it as a simple pest management strategy, but as a program to reduce pesticides by 50 per cent, it was readily accepted." By 1993, six years into the program, pesticide use in Ontario had been reduced 28 percent, and Ontario officials expect the 50 percent reduction goal to be achieved by 1998, when the next pesticide use survey is conducted.

The role of government in reducing pesticide use is crucial, and there are precedents for a heavy-handed and forceful approach. For example, the federal governments of Canada and the United States assign rights to private enterprise that determine commodity harvesting, such as how many fish can be taken or how much lumber cut. Similar rights could be given out for pesticide sales or use, with annual amounts diminishing by 5 percent a year until the desired reduction had been accomplished. This style of mandated pesticide reduction would have some advantages. For one thing, it would accomplish the objective in a clear timeframe and would be effective. Also, pesticide prices would rise as the quantities of pesticide available for sale diminished, so that the cost of chemicals would more accurately reflect their side effects. Finally, increased prices and diminished supplies would encourage the development and use of alternatives.

The chemical industry, farmers, and exterminators, however, would scream at such a heavy governmental hand intervening into their lives and businesses. Moreover, it probably is unnecessary, because there are more creative, positive, and less controversial ways to reduce pesticide use.

Education of professionals who apply pesticides is the most significant role that governments can play to diminish chemical use. Almost every commercial pesticide applicator, whether on a farm or in a city, is required to take some type of course and pass an examination prior to being licensed. This licensing system provides an excellent opportunity to promote a reduction in chemical applications. Pesticide use would gradually diminish if the next generation of pest managers were trained to approach pest problems with a pesticides-last attitude and were provided with sufficient knowledge to implement biologically based technologies. In British Columbia, the new structural pest management manual used in applicator licensing courses already stresses alternative technologies rather than chemical pesticides. A similar approach has been followed in Ontario, where the pesticide education courses required for all farmers are being revised to focus on pesticide reductions for individual crops.

A key element in the success of these two programs has been industry involvement. The applicator licensing programs in British Columbia and Ontario were stimulated by grower and exterminator concerns about pesticide use, and the courses, manuals, recommended procedures, and license examinations were developed in partnership with industry. Government can impose changes in pest management, but the system works best when industry demands changes and has real input into creating the new status quo.

Government also could use its regulatory powers to accelerate the granting of permits for the use of new biologically based pesticides, and retard the adoption of new chemicals. The permit process for alternative technologies could be streamlined by the use of focused technical advisory panels composed of representatives from research, industry, government, and environmental groups. In addition, consolidation of the various agencies and offices involved in regulating pest management would result in more rapid approval for low-risk technologies and shorten the implementation time for alternative technologies. The EPA Pesticides Branch recently began to follow this approach, and now approves most biologically based technologies in a matter of months, whereas chemical pesticide approvals take many years.

Government also can exert influence through its ability to tax. A combination of tax increases on pesticide sales and tax relief for those using alternatives could be a powerful incentive for chemical users to switch to biologically based methods. We frequently subsidize the consequences of individual or commercial actions when government puts taxes on some things to pay for others. For pesticides, users pay for the chemicals they purchase and the costs of application, but government pays for the side effects, including most environmental clean-up costs and some health care expenses. This system needs to be more balanced, and chemical users taxed more heavily to pay for the nontarget impact of pesticide use. Higher pesticide costs through taxation would not be popular among users, but it would make alternative technologies relatively more cost-effective than they are today.

It seems fair for the users of pesticides to pay more directly for the damage they cause, but of course some or all of those cost increases would be passed on to consumers. Slightly higher grocery prices, increased rents for apartment dwellers, and higher tax levies on property owners might be some economic consequences of government using its taxation abilities to discourage pesticide use. However, decreased chemical use would save all of us billions of dollars each year by reducing current government expenditures that are necessary to regulate pesticides and clean up their side effects, although these savings might not be immediately apparent.

❧

It is all well and good to tell government to legislate reduced pesticide use, but that is not enough. We need pest management alternatives that go beyond the conceptual into the realm of the practical and profitable. A final way in which government could intervene is to use its funding power to encourage the research community to go further into the realm of implementation research to bring novel methods to the marketplace.

Simply tossing more money at alternative research is not the answer, however. As we have seen, government already spends hundreds of millions of dollars each year to fund research into biologically based pest management, with little to show for it. To date, too much research has concentrated on inventing alternatives that languish in the twilight zone of interesting but impractical ideas. There has not been enough practical work to bring new concepts to the product stage.

Perhaps an interesting way to encourage a more practical research agenda would be to create a different mix of background and talent on boards that disperse grant funds. Currently, most grants are awarded by peer panels that review applications from fellow scientists. That system has produced interesting science but not enough commercially available products. We need a companion system for applied research, with granting panels composed of users as well as scientists.

The best arrangement would be for a mix of scientists, extension agents, growers, and industrial representatives to decide on funding. Such a mixed group would be more likely to identify projects with a good probability of leading to practical solutions. This is not to diminish the importance of basic research, which still must come before successful applications. Nevertheless, rigorously evaluated funding decisions to encourage more genuinely useful applied research would allow us to benefit from fundamental advances that have the potential to reach the product stage. The recent implementation of small business innovation grants and university–industry liaison programs by the USDA, the National Science Foundation, and the Natural Sciences and Engineering Research Council of Canada are a positive step in that direction.

Another way to foster a more useful research contribution toward alternative technologies is to provide rewards for success in this area. Today, career advancement for university and government scientists is still based primarily on publication quantity and quality in basic research, rather than on patents and industrial contributions. This may be changing, but if so it is changing slowly. Academic scientists would be more encouraged to enter the practical arena if successful product development brought both royalties and promotions. Again, it would be unwise to force our university scientists solely into industrial collaboration, but a better mix of basic research and university–industry interactions would create a healthier and more useful research environment.

The area-wide program for codling moth control being conducted in Washington, Oregon, and California is funded largely by the U.S. Department of Agriculture, but the stakeholders in this project include government and university scientists, state extension agents, private industry, and growers. The program has worked because each group has real input into the project, and the combined experience and perspective of the various parties represented has kept it grounded in real-world economics. The scientists have provided a conceptual basis for the project, and ensured that the applied science being done is rigorous. Extension agents bridge the gap between the university campus and the field. The growers provide practical experience about what works for them in the field and at bill-paying time; since they have real input and share decision-making, the growers have been willing to try new methods. Industry in the form of pheromone-producing companies has been important in working with the team in advising on pricing and product development; in the end, companies must be able to make a profit selling the mating disruption product or the entire enterprise will fail.

Owing to its blend of participants, the codling moth project has not made the common errors of studying an alternative technology in isolation or refusing on principle to use any chemicals. Rather, codling moth control has been considered as only one of many components needed in a holistic pest management program. For example, the reduced chemical sprays that follow a successful mating disruption program can result in increased populations of other pests formerly controlled by pesticides. The program has released various predators or parasites to control these secondary pests in outbreak situations. Pesticides are used only when the alternatives have not worked and one of the apple pests is causing economic damage too serious to ignore.

Perhaps the best evidence of the success of this broad and flexible program is that the acreage being treated with mating disruptants is increasing dramatically each year, in spite of the higher up-front costs involved in using pheromones, and use of the cheaper chemical pesticides has been decreasing. If this approach is effective on codling moth infesting apples, it can work on any pest infesting any crop, kitchen, lawn, garden, or urban back alley.

Another problem in implementing biologically based technologies has been financial. Industry has lost money on alternatives, with a few exceptions such as transgenic plants and *Bacillus thuringiensis*. Pheromones, predators, parasites, and viruses simply have not proven economically viable, and it is not surprising that businesses involved in this field are composed primarily of small niche companies feeding on the fringes of pest management. Larger industries have made a considerable amount of money on pesticides, and have not perceived the potential for similar profits with alternatives.

Options that might stimulate industry to be a more forceful player in the alternative technology field include, for example, congressional funding for more liaison research between government and university laboratories and private enterprise, thereby reducing research costs to industry for the development of new products. Tax credits and small business loans could be used to encourage smaller companies to enter and persist in new markets with biologically based methodologies. The government itself could increase its purchasing and use of alternatives. Governmental agencies are possibly the largest single manager of pests in our society, and the markets created by government buying power would provide considerable sales potential and stability to biologically based industries. Finally, patent protection is vitally important in encouraging industry to enter new markets, and further extending patent rights into biologically based areas would have real impact in stimulating private enterprise to move into alternative technologies.

A final problem with pest management today is that the public continues to object to the implementation of safe alternative technologies. I find this odd, because most North Americans complain about pesticide use in theory, even those who use pesticides in practice. We seem to have developed a generic fear of new technologies and prefer using familiar chemical pesticides even when alternatives have passed rigorous health and environmental safety requirements.

Some environmental groups have become an impediment to the use of nonchemical methods. This is supremely ironic, because environmentalists alerted us to the dangers of chemical pesticides in the first

place. Today, however, negative knee-jerk reactions to new technologies by environmental groups, amplified by the media, have made it difficult for the public to assess the safety of biologically based techniques that could reduce pesticide use.

This problem was brought home to me in a 1996 sequel to the Vancouver gypsy moth "crisis," when a proposed *B.t.* spray in New Westminster, a Vancouver suburb, was criticized by the Society Targeting the Overuse of Pesticides (STOP). In this case, the British Columbia Environmental Appeals Board eventually delayed the spray and proposed instead that Agriculture Canada attempt to eradicate the incipient moth infestation using pheromone-baited traps. This trapping technique was suggested by STOP itself as an alternative to *B.t.* in their submission to the Appeals Board, yet when their pheromone-based proposal was accepted, STOP members began stridently and vociferously objecting to pheromones.

STOP President Christopher Lewis wrote to the local newspaper, the *News Leader,* saying, "What effect does a non-human sex hormone have on our own body, chemistry, hormones, and reproductive systems? And could the offspring of contaminated individuals inherit the insect sex hormone in a bizarre exchange of mutated genetic material . . . as well as causing cancer of the breast, ovaries, or testicles?" What he failed to point out is that gypsy moths themselves release considerably more pheromone than is found in a trapping program, and there is no evidence that the natural pheromone causes any health problems, even in outbreak areas where there may be many millions of individual moths.

In situations like this, the media do not always exercise a reasonable level of discrimination between legitimate concerns of dedicated environmentalists and posturing from uninformed sources. Our society does not, and should not, muzzle the press, and we should encourage environmental advocates to maintain their vigilant scrutiny of new scientific advances. Nevertheless, reporters could do a more responsible job of sifting through real and imagined dangers, and the public needs to be more critical of reports concerning pest management. The press would be a more useful source of information, and contribute more to the dialogue about pest management practices, by highlighting substantive issues and qualified spokespersons rather than misleading claims about biologically based alternatives.

Our perceptions about pest management also make little connection between what we purchase and how those products were treated prior to appearing on store shelves. Perhaps we should label pesticide-treated commodities the way we label cigarette packages, with informative messages about health and environmental risks. Picture a sign above a bin of tomatoes in the grocery store that says, "These tomatoes have been treated with the chemical pesticide Diazinon to control the tomato fruitworm. Diazinon kills beneficial predators and parasites." That would certainly provide an interesting wake-up call to consumers. In the next bin, a similar batch of tomatoes might be labelled with signs informing the consumer that "these tomatoes are pesticide-free, and the tomato fruitworm was controlled using a parasitic wasp that is harmless to human or environmental health." Imagine similar messages on cotton clothing, telling potential buyers that one shirt was woven from cotton on which chemicals had been used to control bollworms, and another shirt labeled as chemical-free, with bollworms controlled by pheromones.

There clearly are many tangible steps we could take as individuals and as a society toward more responsible pest management. Government could use its influence and power to educate, regulate, and tax. The academic community could shift some of its research focus toward more practical ends. Industry and pesticide users could be encouraged through various incentives to explore, market, and use more biologically based technologies. The public could be made more aware of the negative impact that widespread pesticide use has, and of the feasible alternatives that could dramatically reduce our use of synthetic chemical pesticides. None of these ideas would be difficult to implement, and all are affordable.

But before pest management practices change, there is another level of understanding that we much reach, one that goes beyond the economic and scientific facts that have made us pesticide-dependent. That level is an ethical one, where we confront our duty to ourselves, society, and the natural world around us and formulate collective rules of conduct. When it comes to pesticide use, we have not adopted a set of ethics. There is an important ethical difference between protecting ourselves from real pests and indiscriminately attacking innocuous organisms. There also is an important difference between poisoning ourselves and  other nontarget organisms with excessive or improper use of chemical pesticides, and conducting a biologically based pest management program with limited pesticide use and few or no side effects. Our human footprint is heavy on the Earth today, but it would be possible to maintain our health and lifestyles with a lighter step, at least as far as pest management is concerned.

Our war against pests has exacted a heavy toll, but it is a toll we do not see. We have come to expect a sanitized world, a world without cockroaches in our kitchens, insects on our lettuce, dandelions on our lawns, or blemishes on our oranges. That sanitized existence has a price, however, in thousands of serious or fatal pesticide poisoning cases each year, especially in developing countries, and in fields, meadows, and forests stripped of biological diversity. Not only do we annihilate pests, but we also take out plants, insects, fish, birds, and myriad other organisms when we spray.

We should not expect to make peace with pests, or expect them to stop harming us if we embrace them. We can, however, manage rather than control, reduce instead of eradicate, tolerate rather than panic. We can agree on a new pest management ethic, and we have the means and technical ability to implement it. Pest management does not have to be an endless war against nature. It can express the highest ideals of the human spirit to be stewards of our environment while simultaneously taking effective action to protect ourselves from the damage caused by real pests. We have an obligation to ourselves, and a duty to the world around us, to attempt to live according to that ideal.

REFERENCES

ACKNOWLEDGMENTS AND SOURCES

INDEX

References

Numbers in parentheses refer to chapters for which that reference was used.

Adams, L. W. 1994. *Urban Wildlife Habitats.* Minneapolis: Univ. Minnesota Press. (4)

Adams, R. W. 1992. *Handbook for Pesticide Applicators and Dispensers.* Victoria: Pesticide Management Branch, B.C. Ministry of Environment. (3, 4)

Agriculture Canada. 1993. *Information Submitted to the B.C. Environmental Appeal Board.* Vancouver: Agriculture Canada. (2)

______ 1985. *An Evaluation of the Commercial Cost of a Sterile Insect Release Control Program for Codling Moth in British Columbia.* New Westminster, B.C.: Regional Development Branch. (5)

______ 1986. *Analysis of the Risks and Costs of a Sterile Insect Release Program for the Control of Codling Moth in the Okanagan Region of British Columbia.* New Westminster, B.C.: Regional Development Branch. (5)

______ 1994. *Assessment Criteria for Determining Environmental Safety of Plants with Novel Traits.* Regulatory Directive 94–08. Nepean, Ontario: Agriculture Canada. (8)

______ 1996. *Determination of Environmental Safety of NatureMark Potatoes' Colorado Potato Beetle (CPB) Resistant Potato (Solanum tuberosum L.).* Decision Document DD96–06.Nepean, Ontario: Agriculture Canada. (8)

Alcamo, I. E., and A. M. Frishman. 1980. The microbial flora of field-collected cockroaches and other arthropods. *J. Environmental Health* 42:263–266. (3)

Alderson, C., and A. Greene. 1995. Bird deterrence technology for historic buildings. *APT Bulletin* 26:18–30. (4)

Ames, B. N., and L. S. Gold. 1991. Cancer prevention strategies greatly exaggerate risks. *Chemical and Engineering News* 69:28–32. (9)

Ames, B. N., M. Profet, and. L. S. Gold. 1990. Nature's chemicals and synthetic chemicals: comparative toxicology. *Proc. Natl. Acad. Sci. USA* 87:7782–7786. (9)

Anderson, D. J., and R. A. Hites. 1988. Chlorinated pesticides in indoor air. *Environ. Sci. Technol.* 22:717–720. (3)

Anderson, S. H. 1985. *Managing Our Wildlife Resources.* Toronto: E. Merrill Publishing. (4)

ARA Consulting Group, Inc. 1996. *SIR Program Review.* Vancouver, B.C. (5)

Ashton, F. M., and T. J. Monaco. 1991. *Weed Science: Principles and Practices.* Wiley: New York. (1,4)

Aspelin, A. L. 1994, 1996. *Pesticides Industry Sales and Usage.* Washington, D.C.: U.S. Environmental Protection Agency, Office of Pesticide Programs. (1,3,7)

Atkinson, K. T., and E. M. Shackleton. 1991. Coyote, *Canis latrans,* ecology in a rural-urban environment. *Canadian Field-Naturalist* 105:49–54. (4)

Attfield, R. 1983. *The Ethics of Environmental Concern.* Oxford: Blackwell. (Preface)

Banfield, M. G. 1991. An analysis of the semiochemical industry in North America. MBA Thesis, Simon Fraser University. (6)

Bardin, P. G., S. F. van Eeden, J. A. Moolman, A. P. Foden, and J. R. Joubert. 1994. Organophosphate and carbamate poisoning. *Archives Internal Medicine* 154:1433–1441. (3)

Bauer, L. S. 1995. Resistance: a threat to the insecticidal crystal proteins of *Bacillus thuringiensis. Florida Entomologist* 78:414–443. (8)

Beard, J. B., and R. L. Green. 1994. The role of turfgrasses in environmental protection and their benefits to humans. *J. Environ. Quality* 23:452–460. (4)

Berenbaum, M. R. 1995. *Bugs in the System.* Reading, Mass.: Addison-Wesley. (1,3,7)

Bloem, K. A., and S. Bloem. 1995. Codling moth eradication program in British Columbia: a review and update.*Proc. International Cherry Fruit Fly Symposium.* Corvallis: Oregon State University Agricultural Communications. (5)

Bolen, E. G., and W. L. Robinson. 1995. *Wildlife Ecology and Management,* 3rd ed. New York: Prentice Hall. (4)

Brahams, D. 1994. Lindane exposure and aplastic anaemia. *The Lancet* 343:1092. (3)

Brenner, R. J., P. G. Koehler, and R. S. Patterson. 1987. Health implications of cockroach infestations. *Infections in Medicine* 4:349–360. (3)

Buckley, P. A., and M. G. McCarthy. 1994. Insects, vegetation, and the control of laughing gulls (*Larus atricilla*) at Kennedy International Airport, New York City. *J. Applied Ecology* 31:291–302. (4)

Burger, J. 1985. Factors affecting bird strikes on aircraft at a coastal airport. *Biological Conservation* 33:1–28. (4)

Butler, R. W., and R. W. Campbell. 1987. The birds of the Fraser River delta: populations, ecology and international significance. Occasional Paper Number 65. Delta, B.C.: Canadian Wildlife Service. (4)

Butt, B. A., J. F. Howell, H. R. Moffitt, and A. E. Clift. 1972. Suppression of populations of codling moths by integrated control (sanitation and insecticides) in preparation for sterile-moth release. *J. Economic Entomology* 65:411–414. (5)

Butt, B. A., L. D. White, H. R. Moffitt, D. O. Hathaway, and L. G. Schoenleber. 1973. Integration of sanitation, insecticides, and sterile moth releases for suppression of populations of codling moths in the Wenas Valley of Washington. *Environmental Entomology* 2:208–212. (5)

Campbell, S. 1995. *Naturescape British Columbia: Caring for Wildlife Habitats at Home.* Victoria, B.C.: Ministry of Environment, Lands, and Parks, Province of British Columbia. (4)

Carbyn, L. N. 1989. Coyote attacks on children in western North America. *Wildl. Soc. Bull.* 17:444–446. (4)

Carson, R. 1962. *Silent Spring.* Boston: Houghton-Mifflin. (1,7,9,10)

Cartmill, M. A. 1993. *View to a Death in the Morning.* Cambridge: Harvard University Press. (Preface)

Cockroach Control (undated). The Best of PCT. *Pest Control Technology.* (3)

Cram, W. A. 1989. *Gaining Support for British Columbia's Gypsy Moth Wars, 1978–1988.* B.C. Pest Management Report No. 12. Vancouver, B.C.: Ministry of Forests. (2)

Debach, P., and D. Rosen. 1991. *Biological Control by Natural Enemies.* Cambridge: Cambridge University Press. (7)

DeBiasio, D. 1988. Codling moth sterile insect release study. Kelowna, B.C.: B.C. Fruit Growers Association. (5)

Deschamps, D., F. Questel, F. J. Baud, P. Gervais, and S. Dally. 1994. Persistent asthma after acute inhalation of organophosphate insecticide. *The Lancet* 344:1712. (3)

Djerassi, C., C. Shih-Coleman, and J. Diekman. 1974. Insect control of the future: operational and policy aspects. *Science* 186:596–607. (7)

Doane, C. C., and M. L. McManus. 1981. *The Gypsy Moth: Research toward Integrated Pest Management.* Technical Bulletin 1584. Washington, D.C.: U.S. Dept. Agriculture Forest Service Science and Education Agency. (2)

Doe, J. E., and G. M. Paddle. 1994. The evaluation of carcinogenic risk to humans: occupational exposures in the spraying and application of insecticides. *Regulatory Toxicology and Pharmacology* 19:297–308. (3)

Dolstad, K. D. 1985. *Biology and Control of the Codling Moth in the Pacific Northwest.* MPM Thesis, Simon Fraser University. (5,6)

Dubos, R. 1986. The Wilderness Experience. In *People, Penguins, and Plastic Trees,* ed. D. Van DeVeer and C. Pierce. Belmont, Cal.: Wadsworth. (Preface)

Dyck, V. A., S. H. Graham, and K. A. Bloem. 1992. Implementation of the sterile insect release programme to eradicate the codling moth in British Columbia, Canada. *Proc. International Symposium on Management of Insect Pests.* Vienna: FAO, United Nations. (5)

Ecogen, Inc. 1996. 10-K Submission. Langhorne, Pa. (6)

Ellstrand, N. C. 1988. Pollen as a vehicle for the escape of engineered genes. *Trends in Ecology and Evolution* 3:30–32. (8)

Ellstrand, N. C., and C. A. Hoffman. 1990. Hybridization as an avenue of escape for engineered genes: strategies for risk reduction. *BioScience* 40:438–442.(8)

Entwistle, P. F., J. S. Cory, M. J. Baily, and S. Higgs. 1993. *Bacillus thuringiensis, an Environmental Biopesticide: Theory and Practice.* New York.: J. Wiley and Sons. (2,8,9)

Environment Canada. 1993. *A Comprehensive Survey of Pesticide Use in British Columbia.* Pesticide Management Program Publication 93–3. Victoria, B.C.: Ministry of Environment, Lands, and Parks. (3,4)

Fabre, J. H. 1914. *Bilder aus der Insektenwelt.* Stuttgart: Verlag Kosmos. (6)

Fenske, R. A., K. G. Black, K. P. Elkner, C. Lee, M. M. Methner, and R. Soto. 1990.

Potential exposure and health risks of infants following indoor residential pesticide applications. *Amer. J. Public Health* 80:689–693. (3)

Fincham, J. R. S., and J. R. Ravetz. 1991. *Genetically Engineered Organisms: Benefits and Risks.* Buckingham, U.K.: Open University Press. (8)

Flint, M. L., and R. van den Bosch. 1981. *Introduction to Integrated Pest Management.* New York: Plenum Press. (1,7,9)

Forbush, E. H., and C. H. Fernald. 1896. *The Gypsy Moth.* Boston: Wright and Potter Printing Co. (2)

Fox, J. L. 1994. Do transgenic crops pose ecological risks? *Bio/Technology* 12:127–128. (8)

Fox, M. W. 1992. *Superpigs and Wondercorn.* New York: Lyons and Burford. (8)

Fox, W. J. 1994. *Review and Assessment of Times and Practices Recommended for Reentry into Buildings after Application of Residual Insecticides.* Masters thesis, Simon Fraser University. (3)

Free, J. B. 1993. *Insect Pollination of Crops.* New York: Academic Press. (7)

Gerardi, M. H., and J. K. Grimm. 1979. *The History, Biology, Damage, and Control of the Gypsy Moth Porthetria dispar L.* Cranbury, N.J.: Associated University Presses. (2)

Gilkeson, L. A., and R. W. Adams. 1996. *Integrated Pest Management Manual for Structural Pests in British Columbia.* Victoria, B.C.: British Columbia Ministry of Environment, Lands, and Parks. (3)

Gill, D., and P. Bonnett 1973. *Nature in the Urban Landscape.* Baltimore: York Press. (4)

Glacken, C. J. 1967. *Traces on the Rhodian Shore.* Berkeley: University of California Press. (Preface)

Glaser, V. 1995. Pheromone firms cooperate to speed approvals. *Bio/Technology* 13:219–220. (6)

Goldburg, R. J., and G. Tjaden 1990. Are B.t.k. plants really safe to eat? *Bio/Technology* 8:1011–1015. (8)

Goldburg, R., J. Rissler, H. Shand, and C. Hassebrook. 1990. *Biotechnology's Bitter Harvest.* Cambridge, Mass.: Biotechnology Working Group. (8)

Gould, F. 1988. Evolutionary biology and genetically engineered crops. *BioScience* 38:26–33. (8)

Gould, F. 1991. The evolutionary potential of crop pests. *American Scientist* 79:496–507. (8)

Goy, P. A., and J. H. Duesing. 1995. From pots to plots: genetically modified plants on trial. *Bio/Technology* 13:454–458. (8)

Goy, P. A., E. Chasseray, and J. Duesing. 1994. Field trials of transgenic plants: an overview. *Agro-Food Industry Hi-tech* 5: 10–15. (8)

Graham, F. 1970. *Since Silent Spring.* Houghton Mifflin: Boston. (1,9)

Graham, J. D., and J. B. Wiener. 1995. *Risk vs. Risk: Tradeoffs in Protecting Health and the Environment.* Cambridge: Harvard Univ. Press.

Greene, A. 1992. Terminating exterminating. *Federal Managers Quarterly* 4:8–12. (3)

Greene, A. 1996. Pest control turns green. *Forum for Applied Research and Public Policy* 11:76–80. (3,9)

Grierson, D. 1991. *Plant Genetic Engineering.* London: Blackie. (8)

Gut, L. J., and J. F. Brunner. 1994. Pheromone-mediated control of codling moth in Washington apple orchards. *Good Fruit Grower* 45:35–48. (6)

Hall, R. W., and L. E. Ehler. 1979. Rate of establishment of natural enemies in classical biological control. *Bull. Entomol. Soc. Amer.* 25:280–282. (7)

Hall, R. W., L. E. Ehler, and B. Bisabri-Ershadi. 1980. Rate of success in classical biological control of arthropods. *Bull. Entomol. Soc. Amer.* 26:111–114. (7)

Hecker, E., and A. Butenandt. 1984. Bombykol revisited-reflections on a pioneering period and on some of its consequences. In *Techniques in Pheromone Research,* ed. H. E. Hummel and T. A. Miller. New York: Springer-Verlag. (6)

Health Action. 1993. "Jurassic Park" in our supermarkets. *Health Action,* October, pp. 1–3 (8)

Hewitt, T. I., and K. R. Smith. 1995. Intensive agriculture and environmental quality: examining the newest agricultural myth. *Report from the Henry A. Wallace Institute for Alternative Agriculture, Greenbelt, Maryland,* September 1995, pp. 1–12. (1,9)

Hiatt, A. 1993. *Transgenic Plants.* New York: Marcel Dekker.(8)

Howell, R. G. 1982. The urban coyote problem in Los Angeles County. *Proc. Tenth Vertebrate Pest Conference* 10:21–23. (4)

Hoyle, R. 1994. A quixotic assault on transgenic plants. *Bio/Technology* 12:236–237. (8)

______ 1995. EPA okays first pesticidal transgenic plants. *Bio/Technology* 13:434–435. (8)

______ 1996. How not to take the risk out of transgenic plants. *Bio/Technology* 14:142–143. (8)

Hynes, H. P. 1989. *The Recurring Silent Spring.* New York: Pergamon Press. (9)

James, M. T., and R. F. Harwood. 1969. *Herms's Medical Entomology.* London: Macmillan. (1)

Jones, D. D. 1994. *The Biosafety Results of Field Tests of Genetically Modified Plants and Microorganisms.* Oakland: University of California Press. (8)

Judd, G. J. R., and M. G. T. Gardiner. 1993. Pheromone-based control of codling moth: successful and registered. *British Columbia Orchardist* 33:10–11. (6)

Julien, M. H., J. D. Kern, and R. R. Chan. 1984. Biological control of weeds: an evaluation. *Protection Entomology* 7:3–25. (6)

Kaiser, J. 1996. Pests overwhelm *B.t.* cotton crop. *Science* 273:423. (8)

Knight, A. L. 1995. What do we know about the use of sex pheromones? *Good Fruit Grower* 46:37–54. (6)

______ 1995. The impact of codling moth (Lepidoptera: Tortricidae) mating disruption on apple pest management in Yakima Valley, Washington. *J. Entomol. Soc. Brit. Columbia* 92:29–38. (6)

Krattiger, A. F., and A. Rosemarin. 1994. *Biosafety for Sustainable Agriculture.* International Service for the Acquisition of Agri-biotech Applications. Stockholm: Stockholm Environment Institute. (8)

Krimsky, S. 1991. *Biotechnics and Society: The Rise of Industrial Genetics.* New York: Praeger. (8)

Kung, S., and R. Wu. 1993. *Transgenic Plants.* New York: Academic Press. (8)

Kydonieus, A. F., and M. Beroza. *Insect Suppression with Controlled Release Pheromone Systems.* Boca Raton, Florida: CRC Press. (6)

Leiss, W. 1994. *The Domination of Nature.* Montreal: McGill-Queens University Press. (Preface)

Leiss, J. K., and D. A. Savitz. 1995. Home pesticide use and childhood cancer: a case-control study. *Amer. J. Public Health* 85:249–252. (3)

Lopez-Carillo, L., and M. Lopez-Cervantes. 1993. Effect of exposure to organophosphate pesticides on serum cholinesterase levels. *Archives of Environmental Health* 48:359–363. (3)

MacCracken, J. G. 1982. Coyote foods in a southern California suburb. *Wildl. Soc. Bull.* 10:280–281. (4)

MacDonald, J. F. 1989. *Biotechnology and Sustainable Agriculture: Policy Alternatives.* Ithaca, N.Y.:National Agricultural Biotechnology Council, Boyce Thompson Institute for Plant Research. (8)

MacDonald, J. F. 1990. *Agricultural Biotechnology: Food Safety and Nutritional Quality for the Consumer.* Ithaca, N.Y.: National Agricultural Biotechnology Council, Boyce Thompson Institute for Plant Research. (8)

——— 1991. *Agricultural Biotechnology at the Crossroads: Biological, Social, and Institutional Concerns.* Ithaca, N.Y.: National Agricultural Biotechnology Council, Boyce Thompson Institute for Plant Research. (8)

Mackauer, M., L. E. Ehler, and J. Roland. 1990. *Critical Issues in Biological Control.* Andover, U.K.: Intercept. (7)

McGaughey, W. H., and M. E. Whalon 1992. Managing insect resistance to *Bacillus thuringiensis* toxins. *Science* 258:1451–1455. (8)

McKibben, W. 1989. *The End of Nature.* New York: Anchor Books. (Preface)

Mallis, A. 1990. *Handbook of Pest Control,* 7th ed. Cleveland: Franzak and Foster. (3,4)

Marx, J. L. 1989. *A Revolution in Biotechnology.* Cambridge: Cambridge University Press. (8)

Massachusetts Audobon Society. 1995. Green Deserts. *Sanctuary* 34:3–21. (4)

Mausberg, B., and M. Press-Merkur. 1995. *The Citizen's Guide to Biotechnology.* Ottawa: Canadian Institute for Environmental Law and Policy. (8)

Mearns, J., J. Dunn, and P. R. Lees-Haley. 1994. Psychological effects of organophosphate pesticides: a review and call for research by psychologists. *J. Clinical Psychology* 50:286–294. (3)

Mellon, M., and J. Rissler. 1995. Transgenic crops: USDA data on small-scale tests contribute little to commercial risk assessment. *Bio/Technology* 13:96. (8)

Mennear, J. H. 1994. Dichlorvos carcinogenicity: an assessment of the weight of experimental evidence. *Regulatory Toxicology and Pharmacology* 20:354–361. (3)

Merchant, C. 1980. *The Death of Nature: Women, Ecology, and the Scientific Revolution.* New York: Harper and Row. (Preface)

Metcalf, R. L., and W. Luckmann. 1975. *Introduction to Insect Pest Management.* New York: Wiley and Sons. (1,7)

Mikkelsen, T. R., B. Andersen, and R. B. Jorgensen. 1996. The risk of transgene spread. *Nature* 380:31. (8)

Mitchell, E. R. 1981. *Management of Insect Pests with Semiochemicals.* New York: Plenum Press. (6)

Nabhan, G. P., and S. L. Buchmann 1996. *The Forgotten Pollinators*. New York: Island Press. (7)

Nash, R. F. 1989. *The Rights of Nature*. Madison: University of Wisconsin Press. (9)

National Academy of Sciences. 1989. *Alternative Agriculture*. Washington, D.C.: National Academy of Sciences. (10)

National Research Council. 1989. *Field Testing Genetically Modified Organisms: Framework for Decisions*. Washington, D.C.: National Academy Press. (8)

Nestle, M. 1996. Allergies to transgenic foods: questions of policy. *New England J. Medicine* 334:726–728. (8)

Nordlee, J. A., S. L. Taylor, J. A. Townsend, L. A. Thomas, and R. K. Bush. 1996. Identification of a brazil-nut allergen in transgenic soybeans. *New England J. Medicine* 334:688–692. (8)

Oelschlaeger, M. 1991. *The Idea of Wilderness*. New Haven: Yale University Press. (Preface)

Olkowski, W., H. Olkowski, and S. Daar. 1991. What is integrated pest management? *The Integrated Pest Management Practitioner* 13:1–7. (7,9)

Organisation for Economic Co-operation and Development. 1993. *Field Releases of Transgenic Plants, 1986–1992: An Analysis*. Paris: Commission of European Communities, Organisation for Economic Co-operation and Development. (8)

Passmore, J. 1980. *Man's Responsibility for Nature*. London: Duckworth and Co. (Preface)

Pedigo, L. P. 1989. *Entomology and Pest Management*. New York: Macmillan. (1,6,7,10)

Perkins, J. H. 1982. *Insects, Experts, and the Insecticide Crisis*. New York: Plenum Press. (9)

Pimentel, D., and H. Lehman. 1993. *The Pesticide Question*. New York: Chapman and Hall. (1,7,9,10)

Proverbs, M. D., J. R. Newton, and D. M. Logan. 1977. Codling moth control by the sterility method in twenty-one British Columbia orchards. *J. Economic Entomology* 70:667–671. (5)

Proverbs, M. D., J. R. Newton, and C. J. Campbell. 1982. Codling moth: a pilot program of control by sterile insect release in British Columbia. *Canadian Entomologist* 114:363–376. (5)

Quarles, W. 1994. Mating disruption for codling moth control. *The Integrated Pest Management Practitioner* 16:1–12. (6)

Reeves, J. D., D. A. Driggers, and V. A. Kiley. 1981. Household insecticide associated aplastic anaemia and acute leukaemia in children. *The Lancet* 323:300–301. (3)

Regan, T. 1984. *Earthbound: Introductory Essays in Environmental Ethics*. Prospect Heights, Ill.: Waveland Press. (Preface)

Riley, C. V. 1893. Parasitic and predaceous insects in applied entomology. *Insect Life* 6:130–141. (7)

Rissler, J., and M. Mellon 1993. *Perils amidst the Promise*. Cambridge, Mass.: Union of Concerned Scientists. (8)

Robinson, W. H. 1996. Integrated pest management in the urban environment. *American Entomologist* 42:76–78. (3)

Robinson, W. S., R. Nowogrodzki, and J. M. Williams. 1989. The value of honey bees as pollinators of U.S. crops. *American Bee Journal* 129:411–423. (7)

Roelofs, W. L., A. Comeau, A. Hill, and G. Milicevic. 1971. Sex attractant of the codling moth: characterization with electroantennogram technique. *Science* 174:297–299. (6)

Shore, W. H. 1994. *The Nature of Nature.* New York: Harcourt Brace. (Preface)

Simonich, S. L., and R. A. Hites. 1995. Global distribution of persistent organochlorine compounds. *Science* 269:1851–1854. (9)

Smith, R. F., T. E. Mittler, and C. N. Smith. 1973. *History of Entomology.* Palo Alto, Cal.: Annual Reviews. (1)

Snetsinger, R. 1983. *The Ratcatcher's Child.* Cleveland: Franzak and Foster. (1,3)

Solman, V. E. F. 1973. Birds and aircraft. *Biological Conservation* 5:79–86. (4)

——— 1981. Birds and aviation. *Environmental Conservation* 8:45–51. (4)

Steinhart, P. 1995. *The Company of Wolves.* New York: Knopf. (Preface, 10)

Sultatos, L. G. 1994. Mammalian toxicology of organophosphorus pesticides. *J. Toxicology and Environmental Health* 43:271–289. (3)

Surgeoner, G. A., and W. Roberts. 1993. Reducing pesticide use by 50% in the province of Ontario: challenges and progress. In *The Pesticide Question,* eds. D. Pimentel and H. Lehman. New York: Chapman and Hall. (10)

Timmons, A. M., Y. M. Charters, J. W. Crawford, D. Burn, S. E. Scott, S. J. Dubbels, N. J. Wilson, A. Robertson, E. T. O'Brien, G. R. Squire, and M. J. Wilkinson. 1996. Risks from transgenic crops. *Nature* 380:487. (8)

Thomas, K. 1983. *Man and the Natural World.* London: Penguin. (Preface)

Union of Concerned Scientists. 1995. Experimental releases of genetically engineered organisms. *Gene Exchange* 6:16. (8)

U.S. Congress, Office of Technology Assessment. 1979. *Pest Management Strategies.* Washington, D.C.: OTA. (10)

——— 1988. *New Developments in Biotechnology—Field-testing Engineered Organisms: Genetic and Ecological Issues.* OTA-BA-350. Washington, D.C.: OTA. (8)

——— 1995. *Biologically Based Technologies for Pest Control.* OTA-ENV-636. Washington, D.C.: OTA (6,7,9,10)

U.S. Environmental Protection Agency. 1992. *Pesticide Reregistration.* EPA 700-K92–004. Washington, D.C.: EPA. (9)

——— 1994. *Status of Pesticides in Reregistration and Special Review.* EPA 738-R-94–008. Washington, D.C.: EPA. (9)

——— 1994. *Pesticides Industry Sales and Usage.* EPA 733-K-94–001. Washington, D.C.: EPA. (1,9)

——— 1995. *Office of Pesticide Programs Annual Report for 1995.* EPA 730-R-95–002. Washington, D.C.: EPA. (9)

——— 1995. *Partners of Pesticide Environmental Stewardship.* EPA 730-F-95–002. Washington, D.C.: EPA. (9)

———1995. *Citizen's Guide to Pest Control and Pesticide Safety.* EPA 730-K-95–001. Washington, D.C.: EPA. (9)

van den Bosch, R., P. S. Messenger, and A. P. Gutierrez. 1982. *An Introduction to Biological Control.* New York: Plenum Press. (1,7) van der Geest, L. P. S., and H. H. Evenhuis. 1991. *Tortricid Pests.* Amsterdam: Elsevier Science Publishers. (5,6)

VanDeVeer, D., and C. Pierce. 1986. *People, Penguins, and Plastic Trees.* Belmont, Cal.: Wadsworth Publishing. (Preface)

Wallner, W. E., and K. A. McManus. 1989. *Lymantriidae: A Comparison of Features of New and Old World Tussock Moths.* General Technical Report NE-123. Hamden, Conn.: U.S. Dept. Agriculture Northeastern Forest Experiment Station. (2)

Walsh, B. D. 1866. Untitled. *Practical Entomologist,* September 29. (7)

White, L. 1967. The historical roots of our ecologic crisis. *Science* 155:1203–1207. (Preface)

White, L. D., B. A. Butt, H. R. Moffitt, R. B. Hutt, R. G. Winterfeld, L. G. Schoenleber, and D. O. Hathaway. 1972. Codling moths: suppression of populations from releases of sterile insects in the Wenas Valley of Washington. *J. Economic Entomology* 69:319–323. (5)

Whitmore, R. W., J. E. Kelly, and P. L. Reading. 1992. *National Home and Garden Pesticide Use Survey.*Washington, D.C.: U.S. Environmental Protection Agency. (3)

Wickelgren, I. 1989. Please pass the genes. *Science News* 136:120–122. (8)

Wildavsky, A. 1995. *But Is It True? A Citizen's Guide to Environmental Health and Safety Issues.* Cambridge: Harvard University Press. (9)

Wilkie, T. 1996. Sources in science: who can we trust? *The Lancet* 347:1308–1311. (10)

Williams, T. T. 1992. The spirit of Rachel Carson. *Audobon* 1992:104–107. (9)

Williamson, E. R., R. J. Folwell, A. Knight, and J. F. Howell. 1994. Economic analysis of codling moth control alternatives in apple orchards. Farm Business Management Reports EB1789, Cooperative Extension, Washington State University (6)

Wilson, C. L., and C. L. Graham. 1983. *Exotic Plant Pests and North American Agriculture.* New York: Academic Press. (1,7)

Winston, M. L. 1993. Where's the Buzz? *Orion* 12:26–35. (7)

World Health Organization. 1989. *Public Health Impact of Pesticides Used in Agriculture.* Geneva: World Health Organization/United Nations Environment Programme. (1,10)

Wrubel, R. P., S. Krimsky, and R. E. Wetzler. 1992. Field testing transgenic plants. *BioScience* 42:280–288. (8)

Zwiener, R. J., and C. M. Ginsburg. 1988. Organophosphate and carbamate poisoning in infants and children. *Pediatrics* 81:121–126. (3)

Acknowledgments and Sources

I am grateful to Michael Fisher of Harvard University Press for bringing his considerable enthusiasm to this project and for being highly engaged with the writing of this book at every stage. His breadth as a reader pointed me to many sources I had not considered, and our too-infrequent talks about books and authors were a delightful respite from the tedium of writing. I would also like to thank Susan Wallace Boehmer, whose incisive editing substantially improved the manuscript.

I also appreciate the time and effort so many scientists, pest managers, extension agents, farmers, and government officials spent in answering my queries and providing background information about what they did and how they thought about it. I came away from my interviews and other contacts with a high esteem for those working in this complex and controversial field. More specifically, and by chapter, I would like to thank the following:

Chapter 1: Michael Smirle of Agriculture Canada and Russell Nicholson of Simon Fraser University (SFU) kindly corrected my pesticide chemistry, and Arnold Aspelin of the U.S. Environmental Protection Agency provided updated statistics concerning pesticide use in the United States.

Chapter 2: Jon Bell, Barb Edwards, and Ken Wong of Agriculture Canada and Nancy Argyle of the British Columbia Ministry of Forests spent long hours talking with me about the Vancouver gypsy moth program. Judy Myers from the University of British Columbia and Alan Cameron of Pennsylvania State University were sources of considerable background information about gypsy moth biology and management. Rob Grauer from the Economics Department at SFU assisted with some of the calculations presented in this chapter, and Keith Slessor from SFU's Chemistry Department made useful comments on the first draft of the manuscript.

Chapter 3: The following structural pest managers were extraordinarily open about the nature of their business, and each of them spent many hours talking to me about the extermination industry: Ross Bird, David Buchanan, Ursula Dole, Verne Gilpin, Dave Keegan, Harry Linindoll, Bob Lucy, and Allan Vaudry. Linda Gilkeson of the British Columbia Ministry of the Environment also enthusiastically shared her unusual insight into pest control. Angelo Kouris from the Vancouver Health Department told me all about urban pests, and Conway Lum from Mandeville Gardens patiently explained to me the business of selling pesticides to home users. Stanley Schuman from the College of Medicine at the University of South Carolina pointed out some of the flaws in data concerning carcinogenicity of pesticides used around the home. I benefited from attending lectures by Dick Heath of the B.C. Ministry of Environment and Warren Fox of the Burnaby, B.C., School Board about their pest management jobs. In addition, I appreciate information given me by David Freedman, a physician and friend from Coquitlam, B.C., concerning medical aspects of exposure to pesticides, and by Al Greene of the Government Services Administration in Washington, D.C., about structural pest management in government buildings.

Chapter 4: Bobby Corrigan of Purdue University was the first person I spoke with once I decided to do this book, and his interest in rats was an excellent starting point. I spent an enjoyable day with Roch Grondin from the Vancouver International Airport, watching bird management occur beneath the rumble of landing aircraft. Bob Wick, the Executive Director of the Western Canada Turfgrass Association, and George Kirton of Nutri-Lawn spent many hours talking with me about weeds in grass and what they do to manage them, and Steve Wong from the Vancouver Board of Parks and Recreation gave me a similar perspective from a city employee. Kristine Webber and David Shakleton, both from the University of British Columbia, and Peter Alan, a local trapper, introduced me to the world of urban coyotes. Mike Mackintosh, Chair of the Urban Wildlife Committee, and Liz Thunstrom of the Wildlife Rescue Association of B.C. spoke with me numerous times about the fascinating world of wildlife within the city of Vancouver and surrounding suburbs. In addition, Ron Ydenberg of SFU shared his knowledge of local shorebirds; Susan Campbell from the Naturescape Program discussed that program's objectives with me; Gwen Shrimpton of B.C. Hydro informed me about vegetation management by a public utility; and Colin Bates took me on an interesting tour of rat habitats in the city.

Chapter 5: Linda Edwards, an apple grower and consultant who owns and operates Integrated Crop Management in the Okanagan Valley region of British Columbia, offered warm hospitality and a stream of useful insights during a trip to the valley. The time and honesty afforded me by Ken Bloem, Lance Fielding, and Fred Peters of the Sterile Insect Release Program were invaluable in preparing this chapter. Don Thomson from Pacific Biocontrol, Jay Brunner of Washington State University, Murray Isman of the University of British Columbia, and Wayne Still, an Okanagan apple grower, also offered important perspectives on the SIR program.

Chapter 6: Jay Brunner, Linda Edwards, and Don Thomson shared their views on biological and economic aspects of codling moth mating disruption. Make Banfield was helpful by sending updated statistics about the pheromone business and suggesting a number of leads for further information. Dan Miller of Phero Tech, Inc., kindly discussed the pheromone business at some length. In addition, my understanding of pheromone-based biology and the pheromone industry has benefited greatly from my association with Simon Fraser University's Chemical Ecology Research Group, especially John Borden and Keith Slessor.

Chapter 7: Don Elliot of Applied Bionomics on Vancouver Island, B.C., graciously took a day out from his busy schedule to tell me about the business end of biological control. Manfred Mackauer from SFU, Maclay Burt of the Association of Biological Control Producers, and Dave Gingrich of Westgro Industries in Delta, B.C., were the source of many helpful leads and suggestions.

Chapter 8: Sharlene Matten and Mike Mendelson of the U.S. Environmental Protection Agency provided massive documents about biotechnology regulation, and Rick Hellmich of the U.S. Department of Agriculture suggested other resources. Fred Gould of North Carolina State University provided manuscripts and other material; many reprints, news clippings, and ideas were organized by Zamir Punja, Director of SFU's Centre for Pest Management; and data on research funding were made available to me by Robert Faust, U.S. Department of Agriculture, National Program Staff.

Chapter 9: Margie Fehrenbach, Janet Anderson, Arnold Aspelin, and Clarence Lewis, all of the U.S. Environmental Protection Agency, introduced me to that organization's philosophy and activities in the pest management field.

My colleagues and students at the Centre for Pest Management, Simon Fraser University, have created a stimulating and provocative environment in which to learn about pests. Funding provided by a Senior Industrial Fellowship from the Natural Sciences and Engineering Research Council of Canada gave me time to conduct some of the research for this book.

Finally, I am deeply grateful to my wife, Susan Katz, who read every word, listened to endless ideas, and always knew just what to say. It is Susan, and our daughter Devora, whose encouragement and support made this book possible.

Index